# Augmented Reality

# Augmented Reality
## Placing Artificial Objects in Real Scenes
Proceedings of IWAR '98

Edited by

Reinhold Behringer

Gudrun Klinker

David W. Mizell

**CRC Press**
Taylor & Francis Group
Boca Raton  London  New York

CRC Press is an imprint of the
Taylor & Francis Group, an **informa** business

AN A K PETERS BOOK

CRC Press
Taylor & Francis Group
6000 Broken Sound Parkway NW, Suite 300
Boca Raton, FL 33487-2742

First issued in hardback 2019

ISBN 13: 978-1-56881-098-0 (hbk)

Visit the Taylor & Francis Web site at
http://www.taylorandfrancis.com

and the CRC Press Web site at
http://www.crcpress.com

**Library of Congress Cataloging-in-Publication Data**

IWAR '98 (1998 : San Francisco, Calif.)
 Augmented reality : placing artificial objects in real scenes :
proceedings of IWAR '98 / edited by Reinhold Behringer, Gudrun
Klinker, David Mizell.
  p.  cm.
 Includes bibliographical references.
 ISBN 1-56881-098-9
 1. Computer graphics--Congresses. 2. Virtual reality--Congresses.
3. Visualization--Congresses.  I. Behringer, Reinhold.
II. Klinker, Gudrun J. III. Mizell, David. IV. Title.
T385.I94  1998
006--DC21                                  99-17305
                                            CIP

# Contents

# Preface

In the spring of 1998, Reinhold Behringer asked Gudrun Klinker and me to team with him to organize an international meeting of the people doing research on augmented reality (AR). This is a technology which, in concept, dates back at least to the 1960s. It acquired a new focus and impetus in the early 1990s, when people began thinking about how to combine the fast, battery-powered computers that the microelectronics revolution had brought into being, with real-time graphics-rendering techniques and six-degree-of-freedom head trackers in order to visually superimpose computer-generated graphics and text above specific points in the user's view of the real world.

Throughout the 1990s, interest in the technology broadened — along with its definition. Researchers began including sound among the types of computer-generated information that AR could stabilize against the user's real surroundings. Computing research groups throughout North America, Europe, and Asia began investigating AR.

Reinhold, Gudrun and I believe that, by 1998, AR had become a viable, accepted and consequential human-computer interface research topic, one with significant practical applications. The quality of the researchers who participated in the first International Workshop on Augmented Reality, and the ingeniousness and importance of their projects, confirmed our belief.

We wish to thank several organizations and individuals who helped us hold the workshop:

- Reinhold's, Gudrun's and my parent organizations at the time, who each provided financial support: Rockwell Science Center, Fraunhofer, and Boeing.

- The User Interface Software and Technology conference — UIST '98, and its general chair Beth Mynatt, who graciously allowed us to co-locate with them.

- The Army Research Laboratory, who agreed to fund the time Reinhold spent preparing for the workshop.

- The IEEE Computer Society's Task Force on Wearable Information Systems and Task Force on Human Centered Information Systems, who co-sponsored the workshop.

I also want to extend my personal thanks to Reinhold and Gudrun for the energy, effort, and intelligence they contributed to the planning process. Because of them, the first IWAR was a success — as one can see by reading this proceedings volume.

David Mizell, General Chair
Bellevue, Washington, USA
5/25/99

# International Workshop on Augmented Reality 1998 — Overview and Summary

Reinhold Behringer, Gudrun Klinker, and David Mizell

## 1 Introduction

The amount of information available to humankind today is increasing steadily and exponentially. Never before has any individual had as much information at his/her fingertips as the common user of modern information technology. In addition to the problems involved in searching and accessing the required information ("data mining"), another important issue is finding the most efficient way to use the information once it has been found. The easy access to huge amounts of data and information has already revolutionized the way in which many professions are being conducted: architects can use digital blueprints to create, on the fly, arbitrary views of buildings that do not yet exist, and physicians have access to their patients' records, including digital Magnetic Resonance Imaging (MRI) and Computer Tomography (CT) data. Since these data are associated with real-world objects (and subjects), the most intuitive way of showing these graphical data would be to place them into the real world: this way, the architect could see the building he designs, integrated into the real environment, and the physician could see the 3D MRI and CT images overlaid directly onto the patient's body.

For this approach to information presentation, the term "augmented reality" (AR) was coined by Tom Caudell [Caudell 92] in the context of a pilot project in which AR technology was used to simplify an industrial manufacturing process in a Boeing airplane factory. Other industrial applications of AR include virtual prototyping, where the product being developed can be visualized in its real-world surroundings and can be tested for design flaws. Besides those applications, AR technology is useful in error diagnostics and maintenance of complex machinery, where status information and instructions can be overlaid directly on the critical machinery component. In the long term, such systems will reduce maintenance cost, because they will be able to replace expensive printed manuals.

Research in AR has emerged from various other research areas such as virtual reality (VR), wearable and ubiquitous computing, and human-computer interaction (HCI). By exploiting technologies and methods developed in the VR domain, AR bridges the gap between virtual reality and the real world, occupying an intermediate position in the reality–virtuality continuum [Milgram 94]. Although AR is often associated with visualization (starting with the first head-mounted display by Sutherland [Sutherland 68]), augmentation can also be provided in the aural domain [Cohen 93] [Mynatt 97]. AR technology provides a means for intuitive information presentation which enhances perceivers' situational awareness and perception of the real world. This enhancement is achieved by placing virtual objects or information cues into the real world. These techniques can also be used to exploit the natural and familiar modalities for human interaction with the environment, e.g., augmenting paper drawings [Mackay 95], a desk environment [Rauterberg 96], or the pen and notebook paradigm [Szalavari 97].

AR is more than a simple integration of existing technologies, however, although several techniques of various research fields, such as computer vision, virtual reality, sound, AI, and ergonomics, are integrated into AR applications, AR has generated a few new research areas–for example, affine projection. Specific requirements of AR in many areas (tracking, interaction between users and objects or information) also trigger new research efforts in those "integration" fields, greatly expanding the scope of those fields.

## 2   Motivation for This Workshop

Because of the interdisciplinary nature of AR research, progress and results in AR have been published at conferences that are associated with the originating research topics, such as VR ([VRAIS] [VRST]), computer graphics ([SIGGRAPH]), human–computer interface/interaction ([CHI], [HCI]), wearable computing ([ISWC]), and medical computer vision ([MRCAS], [CVRMed], [MICCAI]). The AR community has steadily grown, however, and the need for an event exclusively devoted to AR has become evident. Therefore, early in 1998 a small group of AR researchers decided to organize this workshop as a platform for the annual exchange of AR research results and progress reports, as well as an event where the AR community can come together and get acquainted with each other.

## 3   Organization

The general chair of this workshop was David Mizell (Boeing). Co-organizers were Gudrun Klinker (at that time Fraunhofer Society, Graphics Department) and Reinhold Behringer (Rockwell Science Center). The members of the program committee represented research in both academia and industry:

- Ron Azuma (HRL Laboratories)

- Tom Caudell (University of New Mexico)

- Steve Feiner (Columbia University)

- Eric Foxlin (Intersense Inc.)

- Michitaka Hirose (University of Tokyo)

- Adam Janin (University of California at Berkeley)

- Stephane Lavallee (Faculté de Médicine de Grenoble)

- Blair MacIntyre (Georgia Institute of Technology)

- Paul Milgram (University of Toronto)

- Ulrich Neumann (University of Southern California)

- Jannick Rolland (University of Central Florida)

- Andrei State (University of North Carolina at Chapel Hill)

The workshop was organized by the Rockwell Science Center and co-organized by ACM SIGGRAPH, SIGCHI, and Eurographics. It was co-sponsored by the IEEE Computer Society (by the IEEE Computer Society and its Task Forces on Wearable Information Systems and on Human Centered Information Systems). Financial contributions were made by Rockwell, by the Fraunhofer Society, and by Boeing.

The call for papers was published on the Internet in May of 1998. Two types of contributions were requested: full technical research papers and short position statements. These position statements were intended to lead to discussions in which all workshop participants could ask questions and address related AR issues. By the end of July, the program committee began reviewing the submissions, and in the beginning of September of 1998 the program was defined. Although a wide range of topics had been addressed in the call for papers, more than half of the submissions dealt with the topic of registration, and many of these registration papers described computer vision techniques. Therefore, the second half of the workshop was devoted to this topic. The first half was devoted to presentations about AR applications and novel paradigms.

In order to attract a large attendance of researchers in related fields, IWAR '98 was held in conjunction with the annual conference on User Interface Software Technology (UIST) in San Francisco on November 1.

## 4   Statistics

Eight full papers and 11 position statements were accepted for presentation at IWAR '98. In addition, several short late-breaking position statements were accepted for printed distribution. All of these papers are published in these proceedings.

Forty-seven participants had pre-registered for the workshop. There were an additional 16 on-site registrants, which brought the total number of participants up to 63. About half of the participants came from industry labs or non-profit laboratories; the other half is working at universities. There were 42 participants from the USA, 12 from Europe, 8 from Japan, and 1 from Israel.

## 5   Topics and Discussions

The workshop was divided into four sessions:

1. Applications of Augmented Reality – full papers.

2. AR applications and novel user interface paradigms – position statements and discussion.

3. Registration through Computer Vision – full papers.

4. Registration for AR – position statements and discussion.

The papers and position statements from all sessions are included in these proceedings. During the discussion sessions (2,4), the presenters were given only 5 minutes for a brief overview presentation of their research topic. After a block of such position statements, time was allocated for questions and answers. The discussions that resulted are briefly summarized below.

### 5.1   AR Applications and Novel User Interface Paradigms

*What is the cost of future AR systems?*

One participant expressed the opinion that the major cost factor lies in developing the technology, which is most likely to be driven by the cost of VR hardware. The application software only contributes to a minor share of the cost. Under certain circumstances, the cost of AR can actually be less than that of a conventional information system, because one need not pay for the real environment and objects. Eventually, AR will be a commodity item.

*What has to be done to build truly useful AR applications?*

Several participants pointed out that in order to develop usable systems, designers must pay more attention to user issues: they must conduct user studies and listen to potential end users. Research should not focus on tasks that can already be done well without AR. Instead, research should be targeted towards applications where AR can really make a difference, e.g., design and prototyping.

There was a discrepancy between the views of academic and industrial researchers. Whereas the academic community is targeting the application of AR to daily life in a user-centric paradigm, often implemented in a kind of wearable system, industrial AR researchers argued that the current AR applications are strictly task-specific and driven by the requirements for performing a well-defined task. Consequently, these task-specific approaches may not necessarily provide the most intuitive and advanced interaction with the application, but they are targeted towards making the best possible use of current state-of-the-art technology in order to perform this task in a given context.

In general, both sides agreed that technical progress in the display technology (latency, resolution) and in the general capabilities (repeatability, speed) is necessary for developing acceptable AR systems. AR systems should help eliminate human error and must provide means for doing tasks better than can be done with current technology. In order to give users convincing arguments for more advanced technology, not only the display aspect (3D displays) is important, but also the additional capabilities (e.g., networked information flow).

*What will be a killer application of AR, and who will be the potential users?*

This question was also controversial: One notion was that the AR applications will develop their greatest potential in factory environments, for military training, and for maintenance. Others saw AR applications as most suitable for highly trained experts, such as air traffic controllers, video producers, or biologists. A third possible user is the home consumer in the toy and video markets. The papers which were presented actually covered all of these types of AR applications.

The group failed to define a concrete "killer app"; it was noted that so far only one AR system (Boeing's wire harness AR system for airplane manufacturing) is being produced and sold on a commercial basis.

*What combinations of display/interaction technologies are promising?*

The answers ranged from paradigms imitating real-world interaction modes (e.g., electronic paper and ink) to completely new ubiquitous display concepts. Portability and wearability are considered a huge leap forward, even if such characteristics would compromise the precision of the AR system. One of the most important aspects of the technology is the interaction scheme provided to the user: keyboard-less, untethered methods (voice or gesture input) would provide intuitive means for exploring and controlling the information space made accessible by the computer.

*How will AR technology improve in the next five to 10 years?*

This question received various answers, depending on the respondant's field of research. The computer vision community strongly believes in the role of computer vision for registration and information capture. Others believe that developing more advanced sensors will lead to completely different applications of the computer technology, which is currently used mostly to perform "paper-similar" tasks. It is possible that head-mounted displays (HMDs) will soon mature into light and comfortable devices which will not provoke the current objections.

*Are HMDs necessary for AR?*

This question was not asked explicitly, but many comments revealed the ambivalence of many AR researchers towards these visualization devices. Some see HMDs as essential tools for providing immersive visual display, whereas others see them as bulky devices which are not being accepted by users for anything other than very specific tasks. For many applications, handheld devices could also do the job of information display. Whether such devices would still fit under the rubric of *Augmented Reality* is another question. Many researchers expressed the opinion that AR is more than just visualization: augmentation can be done for the aural sense as well as for other senses, and it remains to be seen whether HMDs would be the optimal tool for such augmentation. Even for visual augmentation, other types of displays can provide task- and application-specific information. However, if HMDs become lighter, smaller, and more comfortable to wear, they will be essential for user-centric registered visual augmentation.

*Is there a common vision of AR?*

During the discussions, no single vision of AR unified participants. The AR community does seem to be working in the direction of AR schemes that are wearable, are ubiquitous, involve all the human senses, and link information with our environment. Associating AR with a certain technology is considered a limiting notion.

## 5.2   Registration for AR

*What are the issues for outdoor AR?*

On the hardware side, outdoor applications must be light-weight and must run on low power. The display must be brighter and offer greater contrast. On the software side, registration in a non-constrained environment provides a large challenge. Computer vision, the method generally considered the most suitable for outdoor registration, must advance to such an extent that it can replace or at least complement the conventional registration sensors. The tracking problem is far from being solved: in an outdoor setting, a wide range of situations can cause current tracking systems to fail. Knowledge of the 3D model of the tracked object or scene is important for most visual tracking approaches. Future work should focus on the automatic reconstruction of such 3D models and on incremental learning about the 3D environment.

*Are fiducial markers or natural features preferable for CV-based tracking?*

Many tracking systems rely on fiducial markers. This approach is acceptable to many industrial customers; however, it is not adequate for use in outdoor settings where marking all objects is simply not feasible. Therefore, developers must improve the technology's ability to use natural features rather than man-made fiducials for visual tracking. Mizell presented Klinker's idea of comparing the performance of various tracking systems by staging a "tracker shoot-out" in which the systems have to solve the task of tracking an object (whose CAD model would be provided).

*Should systems provide automated calibration or should they require a human in the loop?*

There were different opinions about whether and to what extent users would be willing to spend time calibrating AR systems (especially the trackers). If the calibration procedures take more time than is saved by the use of AR information presentation, the advantage of AR is questionable. Therefore, self-calibrating tracking systems are very compelling. However, the technology is not yet that mature, and a human may have to substitute for a lack of computer vision capability and calibrate the system manually–until a better technical solution has been developed. Another question is whether "perfect" registration is really necessary for all AR tasks. Since there is a cost associated with registration precision, many users might settle for less stringent registration requirements and still perform satisfactory AR.

*How much training would be required to use an AR system?*

Performing specific tasks with an AR system can actually reduce the training requirements, as has been shown in the Boeing wire bundle AR project, where completely untrained users working <u>with</u> the AR system were able to work faster than trained personnel working <u>without</u> it.

# 6   Summary and Conclusion

This workshop provided a platform for the exchange of research results and an opportunity for the AR researchers to form a community. The discussions showed the wide spectrum of opinions within that community, but also indicated the common grounds of research in this new area. We are looking forward to IWAR '99, which will be held at the Cathedral Hill Hotel in San Francisco October 20 and 21 in conjunction with the Symposium on Wearable Computing (ISWC).

# References

[Azuma 97]      R.T. Azuma. "A Survey of Augmented Reality," *Presence: Tele-operators and Virtual Environments* Vol. 6, No. 4, pp. 355–385, 1997.

[Caudell 92]    T. Caudell and D. Mizell. "Augmented Reality: An Application of Heads-up Display Technology to Manual Manufacturing Processes." *Proceedings of Hawaii International Conference on Systems Sciences*, Maui, Hawaii, IEEE Press, pp. 659–669, January 1992.

[CHI]           *Conference on Computer-Human Interaction (CHI).* Held annually.

[Cohen 93]      M. Cohen, S. Aoki, and N. Koizumi. "Augmented Audio Reality: Telepresence/VR Hybrid Acoustic Environments." *Proceedings of Workshop on Robot and Human Communication*, IEEE Press, Tokyo, Japan, pp. 361–364, November 1993.

[CVRMed]        *IEEE Conference on Computer Vision, Virtual Reality and Robotics in Medicine (CVRMed).* Held every 2–3 years since 1994. CVRMed I was held 1994 in Nice, CVRMed II was held 1997 in Grenoble in conjunction with MRCAS III. http://curie.imag.fr/gmcao/cvrmed-mrcas/index.html.

[HCI]            *Conference on Human–Computer Interface (HCI).* Held annually since 1992.

[ISWC]           *International Symposium on Wearable Computing (ISWC).* Held annually since 1997.

[Mackay 95]      W. Mackay, D. Pagani, L. Faber, B. Inwood, P. Launiainen, L. Brenta, and V. Pouzol. "ARIEL: Augmenting Paper Engineering Drawings." *Proceedings of Conference on Human Factors in Computing Systems (CHI)*, IEEE Press, Denver, CO, pp. 421–422, May 1995.

[MICCAI]         *International Conference on Medical Image Computing and Computer-Assisted Intervention (MICCAI).* Held for the first time in 1998 in Cambridge, MA; formed by the merger of three previous converences: CVRMed, MRCAS, and VBC, http://www.ai.mit.edu/conferences/miccai98.html.

[Milgram 94]     P. Milgram, H. Takemura, A. Utsumi, and F. Kishino. "Augmented Reality: A Class of Displays on the Reality-Virtuality Continuum." *Proceedings of SPIE Conference on Telemanipulator and Telepresence Technologies* SPIE, Boston, MA, pp. 282–292, October 1994.

[MRCAS]          *Conference on Medical Robotics and Computer Assisted Surgery (MRCAS).* Held every few years since 1992. In 1997, MRCAS III was combined with CVRMed II. In 1998, both were run jointly under a new name, MICCAI.

[Mynatt 97]      E.D. Mynatt, M. Back, R. Want, and R. Frederick. "Audio Aura: Light-Weight Audio Augmented Reality." *Proceedings of ACM UIST '97*, ACM, Banff, Canada, pp. 211–212, October 1997.

[Rauterberg 96]  M. Rauterberg, T. Mauch, and R. Stebler. "Digital Playing Desk: A Case Study for Augmented Reality." *Proceedings of IEEE Workshop on Robot and Human Communication*, IEEE Press, Tsukuba, Japan, pp. 410–415, November 1996.

[SIGGRAPH]       *Special Interest Group on Computer Graphics (SIGGRAPH).* Held annually since 1982.

[Sutherland 68]  I.E. Sutherland. "A Head-Mounted Three Dimensional Display." *Proceedings of Fall Joint Computer Conference*, Thompson Books, Washington, DC, pp. 757–764, 1968.

[Szalavari 97]   Z. Szalavari, and M. Gervautz. "The Personal Interaction Panel — A Two-Handed Interface for Augmented Reality." *Proceedings of EUROGRAPHICS '97*, Blackwell Publishers for Eurographics Association, Budapest, Hungary, September 1997.

[VRAIS]          *Virtual Reality Annual International Symposium (VRAIS).* First
                 held in 1993, then annually since 1995. In 1999, this conference
                 will be held under the name "Virtual Reality" (VR).

[VRST]           *Virtual Reality Software Technology (VRST).* Held annually
                 since 1994.

# Part I

# Applications of Augmented Reality

*In recent years, augmented reality has found its way into a range of application demonstrations, indicating a growing interest in AR on the part of those outside the pure research community. Their work has been made possible because the major components of AR have matured into the kind of technology that is usable by non-specialists. The papers in the following section demonstrate the wide range of possible applications of AR technology, including industrial prototyping and manufacturing, medical visualization, and entertainment. These articles document and exemplify the state of the art in the application of the augmented reality paradigm.*

# Case Studies of See-Through Augmentation in Mixed Reality Project

Kiyohide Satoh, Toshikazu Ohshima,
Hiroyuki Yamamoto, and Hideyuki Tamura

**Abstract.** *This paper introduces two case studies of augmented reality (AR) systems which use see-through HMDs. The first case is a collaborative AR system called $AR^2$ Hockey that requires real-time interactive operations, moderate registration accuracy, and a relatively small registration area. The players can hit a virtual puck with physical mallets in a shared physical game field. The other case study is the MR Living Room, where participants can visually simulate the location of virtual furniture and articles in the half-equipped physical living room. The registration is more crucial in this case because for the requirement of visual simulation. The details of registration algorithms implemented are described as well as the system configurations of both systems.*

## 1 Introduction

Most VR systems we have experienced in this decade made it possible for participants to interact with virtual environments which are totally synthesized in computers. As everyone knows, however, the reality in this synthetic world is limited and far from the reality of the physical world. Thus, mixing the real and virtual worlds is a very natural approach for enhancing or augmenting the reality.

We have been participating in the Research Project on Mixed Reality (MR), whose target is to develop technology that merges the real world seamlessly with the virtual world. Mixed Reality Systems Laboratory Inc. was established in January 1997, specifically to carry out the project, and we intend to extend its life until March 2001. Our research topics include display equipment, such as the HMD (head-mounted display) and the 3D display without eyeglasses, as well as software technologies, such as the registration of the two worlds and the system architecture.

MR [Milgram 94] is a concept that covers augmented reality (AR) and augmented virtuality (AV). Although both AR and AV handle the physical

3

world and the virtual one simultaneously, AR is based on the physical world and AV is constructed on the virtual space. There is no other clear distinction, however, between AR and AV, and MR is a "virtual continuum." The geometrical registration between the physical space and the cyberspace is a common problem in MR. The methods for combining both spaces photometrically, however, vary slightly according to the application. This paper introduces two case studies of AR systems with see-through HMDs, using the registration algorithms developed in the first one and a half years of the MR project.

The first case is a collaborative AR system that requires real-time interactive operations, moderate registration accuracy, and a relatively small registration area. The collaborative AR allows multiple participants to share a physical space surrounding them and a virtual space, visually registered with the physical one. They can also communicate with each other through the mixed space. $AR^2$Hockey (Augmented Reality AiR Hockey)[Ohshima 98] is a collaborative AR system in which players can hit a virtual puck with physical mallets in a shared physical game field. Image quality is not a serious factor in this case.

The other case study is the MR Living Room, where participants can visually simulate the location of virtual furniture and articles in the half-equipped physical living room. In this case, we have to render many objects photo-realistically and support a wide registration area so that the participants can walk around and engage in the visual simulation. That is, the registration between the physical space and the virtual space is more crucial in this case than in the $AR^2$Hockey. This system is therefore equipped with a different type of head tracker and optical see-through HMDs with a wider field of view than those used in the former case.

For each system, an appropriate video-rate registration algorithm is implemented, with head trackers and video cameras attached to see-through HMDs. Especially in the MR Living Room, we have proposed a new registration framework for AR systems.

## 2 Collaborative AR

AR research has so far been done mainly on single-user applications [Azuma 97], [Feiner 93], and [Caudell 92]. New applications, especially in the field of human communication, will become possible if multiple participants can share a physical space, and if we can seamlessly integrate a virtual space into the shared physical space [Billinghurst 97]. For example, it becomes possible for multiple people to collaborate on object design in the physical space while exchanging ideas using virtual objects [Rekimoto 96].

We have developed the AR²Hockey system [Ohshima 98] as a case study of the collaborative AR required for human communications. This section describes the configuration of the AR²Hockey. Air hockey is a game in which two players hit a puck with mallets on a table and shoot it into goals. In our AR²Hockey, the puck resides in a virtual space. This simple application presents the following problems for collaborative AR. First, more than two people share a single physical space and a virtual space. Second, since the puck moves fast, rapid response time becomes essential and the synchronization problem must be solved. Third, since the virtual puck is hit by the physical mallets, the positional error and the time lag must be minimized.

Figure 1(a) shows a scene in which two people are playing AR²Hockey and (b) is an image seen through the HMD while the system is operating.

## 2.1   Hardware Components

### HMDs

This system uses optical see-through HMDs [Hoshi 96]. The HMD contains an LCD of 180,000 pixels and two prisms for each eye. One prism is used to lead images displayed on the LCD to the eye. This prism has two off-axial reflective surfaces. To correct the off-axial aberrations, an aspherical surface without rotational symmetry is used in this prism. By attaching the compensation prism to the outside of the prism, we achieve good see-through view. This HMD gives us 34 degrees of horizontal view angle and 22.5 degrees of vertical.

### Trackers

The HMD uses a magnetic sensor (Polhemus' Fastrak) to measure the player's viewpoint. Since this positional and orientational sensor does not have enough accuracy to produce images without notable displacement,

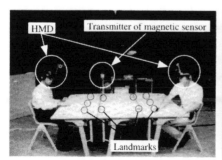

(a) Playing scene          (b) Augmented view (optical see-thru)
**Figure 1.** Playing scene of AR²Hockey

we have placed a small color CCD camera (ELMO) with 45 degrees of view angle near the right eye position of the HMD. This camera detects landmarks in the physical space in order to compensate for the error of the magnetic sensor. See Section 4.1 for the details of the registration algorithm.

Each player holds a physical mallet and uses it to hit a virtual puck. The mallets are simple devices which have infrared LEDs. The positions of the mallets are tracked on the image captured by an infrared CCD camera set directly above the table. In our $AR^2$Hockey, movement of the mallets is constrained to a two-dimensional plane, but 3D tracking may be required depending on applications.

**Computers**

The $AR^2$Hockey system uses three SGI O2s to process two video images from the cameras attached to the HMDs as well as a mallet-tracking process. In addition, one SGI ONYX2 (with eight CPUs and three InfiniteReality graphic pipelines) handles head tracking, image and sound rendering, and total system control. All the computers in the system process in parallel while communicating with each other over the Ethernet network.

By using only one super-graphic workstation to process all the rendering, we can completely synchronize the images given to the participants and thus solve the synchronization problem. Currently, this system distributes processes to its three computers over Ethernet, but it would be ideal to have only one graphics work-station perform all processes, since the time lag for network communication could prove to be a problem.

## 2.2   Process Flow

Figure 2 shows the process flow of this system. Duplicated blocks in the figure represents blocks prepared separately for the two players. As shown in the figure, the process is composed of six types of sub-processes and one main process. Three sub-processes — landmark tracking, registration, and rendering — are invoked for each player.

The four tracking processes that drive input devices work asynchronously. This means that they proceed independently from the main process and parallel to each other. On the other hand, the registration, the space management, and the rendering processes are synchronous with the main process. By configuring the system in this way, we have made it possible to reduce the effect of the differences among the sampling rates of head trackers, the video capturing rate, and the rendering rate. This effect directly influences the time lag of the system.

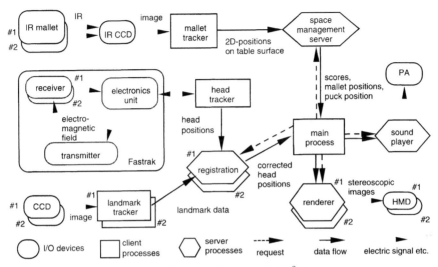

**Figure 2.** Process diagram of AR²Hockey

## Head Tracking Process

The positions of the players' heads are measured by the magnetic sensors attached to the HMDs. The sampling rate is about 50 Hz, depending on the characteristics of the Fastrak system. Note that two sensors for two HMDs are used in the system. The head tracking process receives the head positions at this rate and sends the data to the registration process.

## Landmark Tracking Process

An image as seen from a player's viewpoint is taken from the CCD camera mounted on the HMD. From the image, the system extracts the position of a landmark on the table. The landmark position is sent to the registration process at 30 Hz. However, there is a latency of about 40 ms for capturing and processing an image, so the data sent to the registration process delays for that interval.

Ten small, square-shaped landmarks are placed on the table as shown in Figure1(a). Each landmark is either red or green, depending on which of the two players it is used for.

A landmark is extracted by simple image processing. The process decides that a point is a landmark when it has a certain threshold intensity of the predefined color and appears near the landmark's position from the previous frame. If no such point can be found, the system scans the entire image and detects the point with the nearest color intensity.

### Registration Process

This process registers the images in response to a request from the main process based on the stored head-tracking and landmark-tracking data. Then it sends corrected head-position and orientation data back to the main process.

The latest data stored in this process does not correspond to the current status of the physical world. There is some time delay. Therefore, the registration process records the time stamps that indicate when the data was updated by the tracking processes. Based on the time stamps and the data itself, this process predicts the current head position, orientation, and landmark position using the second-order linear prediction. The registration algorithm is then applied to the predicted data. See Section 4.1 for the registration algorithm.

### Mallet Tracking Process

This process measures 2D positions of mallets on the table. It is implemented by the same simple image processing as we use for the landmark tracker, and it sends mallet-position data to the space management process at 30 Hz with a latency of about 40 ms.

### Space Management Process

This process manages the state of the game, including puck position, speed, and score, in response to the request from the main process, and updates the game status by predicting recent mallet positions based on the stored mallet data. This process also generates sound effects when it updates the game status.

### Rendering Process

This process is synchronous with the main process and generates a set of stereoscopic images. The process obtains the required data from the main process and displays rendered images to the HMD.

### Main Process

This is the process which coordinates and controls the other processes and makes the system highly cohesive. The process requests corrected head-position and orientation data from the registration process, and the game state from the space management process, at the exact rate of 36 Hz. Then it sends these data to the rendering process for rendering.

**Figure 3.** Space of MR Living Room          **Figure 4.** See-through HMD

# 3 Visual Simulation with AR

In the AR$^2$Hockey system, we have studied mainly the static and dynamic registrations, that is, positional misalignment and time lag, while focusing on a game requiring quick motion. "MR Living Room" is another experimental AR system for interior simulation. It was developed using the knowledge we had gained from the AR$^2$Hockey project, but here we took into consideration the technical problems of image quality consistency. This section describes the outline of this project.

The MR Living Room has a 2.8 m × 4.3 m wooden floor half-equipped with a few pieces of furniture and incidental articles. In this space, two participants with see-through HMDs can simulate such activities as selecting and placing furniture. Figure 3 shows the inside of the experiment space. Virtual furniture and articles are merged into this physical space and presented in real time on the HMDs. Figure 5 shows a see-through image and an augmented image.

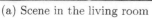

(a) Scene in the living room          (b) Augmented view (video see-thru)

**Figure 5.** Visual simulation in MR Living Room

## HMDs

The HMD in this system uses as its display unit two TFT LCDs of 920,000 pixels; it can present a set of stereo color images with $640 \times 480$ VGA resolution. Though the optical system is basically the same as the HMD used for $AR^2$ Hockey, this HMD has 51 degrees of horizontal view angle and 37 degrees of vertical, with a modified prism that has very little distortion. The transparency rate of the real image is set to a higher value in order to achieve more realistic see-through feeling.

The HMD has a color CCD camera and an optional visor, which shuts out the light from the physical space. When the visor is put on the HMD, the video see-through configuration is realized by displaying the image captured by the color camera. A head tracker and an infrared camera are also mounted for the purposes of registration. Figure 4 shows this HMD.

## Trackers

A hybrid tracker (InterSense's IS-600) is used in the MR Living Room. This tracker uses a gyro-sensor for the orientation measurement and a ultrasonic sensor for the position measurement. The gyro-sensor is suitable for an application in which the participants walk around. However, it shows apparent drift in the heading direction (yaw) because of the measurement error accumulation.

In order to eliminate this kind of measurement error, infrared LEDs are placed in the living room as landmarks and are observed by a small, infrared CCD camera (ELMO's ME411R) mounted on the HMDs. Since the color information is quite important in an application like this interior simulation, it is inappropriate to assign a certain color to the landmarks. Therefore, infrared landmarks are used. The positions of these landmarks on the image are used for registration.

## Computers

This system consists of three graphic workstations (GWS) and two PCs. Two of these GWSs (SGI ONYX2 and O2) are used for image generation, one for each participant. Ideally, two GWSs of ONYX2 class are necessary to present a stereoscopic image to each participant. One O2 is used, however, for one participant because of the limitation of available equipment in our laboratory. Thus, one participant, whose HMD is driven by ONYX2, can see stereoscopic images with optical see-through, but the other can only see monoscopic images with video see-through. A separate GWS (SGI O2) is used for the virtual interior object manipulation and operated by an operator. Two PCs, each of which is equipped with image

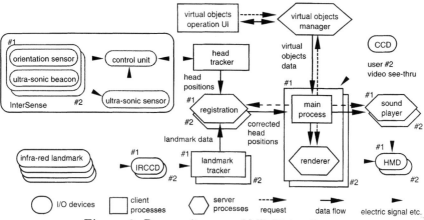

**Figure 6.** Process diagram of MR Living Room

processing hardware (HITACHI IP5005), are used for landmark tracking, one for each participant.

**Process Configuration**

Figure 6 shows the block diagram of this system. The Virtual Interior Object Management Process is the server process which comprehensively manages all information related to the virtual interior objects placed in the virtual space. The operator adds, moves, or deletes the virtual interior objects through the Virtual Interior Object Manipulation process.

# 4 Static Registration Algorithm

For AR systems, the registration of a physical space with a cyberspace is essential. This section describes the registration algorithms implemented in the systems detailed above. The discussion below assumes the perspective projection. All the inner camera parameters are already known and an image is captured without any distortion by an ideal capturing system.

## 4.1 Registration with One Landmark

Correcting positional error using only one landmark is a simple, fast, and effective method for registration. Thus we chose this method in the AR$^2$Hockey system and implemented a registration algorithm based on the method proposed by Bajura et al. [Bajura 95], since the processing speed is crucial for this application.

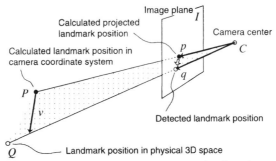

**Figure 7.** Registration with one landmark

Figure 7 shows the basic theory behind correcting positional error using a single landmark. In the figure, let $C$, $I$, and $Q$ be the camera position, the image plane, and the landmark position in the physical 3D space, respectively. For those $C$ and $I$, the landmark $Q$ will be projected and detected on the image at the point $q$ where the line connecting $C$ and $Q$ intersects the image plane $I$. On the other hand, the landmark position $P$ in the camera coordinate system and its projected position $p$ on the image are calculated based on the data from the 3D head tracker. These $P$ and $p$ can be thought of as the landmark in the virtual space and its corresponding image position. Ideally, points $Q$ and $P$ coincide in the 3D space. That is, the projected positions $q$ and $p$ coincide on the image plane. This is, however, usually not true, because of the head tracker's error.

The correction is made by translating the virtual space coordinates so that the calculated projected position of the landmark $p'$ on the image coincides with the point $q$. This is done by translating objects in the virtual space using the equation

$$v = n(\overrightarrow{Cq} - \overrightarrow{Cp}) \tag{1}$$

where $n$ is a scale factor derived by $n = |CP|/|Cp|$ .

Since this method only registers the positions on the 2D image plane, the three-dimensional positions may not coincide correctly even after the registration. This method, however, is still effective if the sensor error is not very great. Moreover, the calculation cost is inexpensive and suitable to the real-time process required by a system like $AR^2$Hockey.

While the collaborative operation requires a broader registration area, the above outlined registration is limited to only a small area. That is because this method requires that a landmark always be observed in the captured images. Placing multiple landmarks in the physical space will make it possible to broaden the registration area. In this case, the problem arises of how to identify the detected landmark. Our method uses the information from the head tracker as a guide to identify the landmark.

Suppose multiple landmarks $Q_i (i = 1 \cdots N)$ are placed in the physical space and their world coordinates are already known. Here $N$ is the number of landmarks. For $Q_i$, let the point $p_i$ denotes the calculated projected image coordinate of the landmark derived from the head tracker. When a landmark is detected at an image coordinate of $q$, we can simply choose the landmark $Q_k$ as the target landmark which minimizes the following evaluation values $e_i$:

$$e_i = |p_i q_i| \quad (i = 1 \cdots N) \tag{2}$$

After the landmark has been decided in this way, the method described above can be used.

## 4.2    Registration with Depth Information

The registration algorithm described in the previous section is very simple, fast, and effective for the AR$^2$Hockey system. The method, however, is not appropriate for the MR Living Room, because the visual simulation application requires higher registration accuracy than that required by AR$^2$Hockey. In addition, the type of information available for registration will change as the participants walk around the space. Thus, we have developed a framework which handles uniformly the various kinds of information available for registration. The following describes this framework.

Assume that a landmark $Q_i$ in the 3D space, whose known world coordinate is $Q_{Wi} = (X_{Wi}, Y_{Wi}, Z_{Wi}, 1)^T$, is projected onto the image coordinate $q_i = (x_i, y_i)$. This projection can expressed by the $3 \times 4$ matrix $\mathbf{C}$ as

$$U_i = \begin{pmatrix} x_i h_i \\ y_i h_i \\ h_i \end{pmatrix} = \mathbf{C} \cdot Q_{Wi} \tag{3}$$

How to obtain this matrix $\mathbf{C}$ is the problem of estimating camera parameter, that is, also the problem of registration. Normally, it is necessary to detect more than six landmarks which are not on the same plane in order to get the matrix $\mathbf{C}$ by linear calculation. In our registration algorithm, even when fewer than six landmarks are detected, it is possible to calculate the camera parameter matrix $\mathbf{C}$ by utilizing the output of the head tracker.

### (1) With four (or more) landmarks

It is known that the parameter $h_i$ in Equation 3 is proportional to the depth value $Z_{Ci}$ of the landmark $Q_i$ in the camera coordinate system [Mellor

95]. If we can measure the depth values for the detected landmarks, three
equations may be derived for each landmark by substituting $h_i$ in Equation
3 with $Z_{Ci}$. If four or more landmarks which are not on the same plane
are detected, Equation 3 can be expressed as

$$\mathbf{U} = \mathbf{C} \cdot \mathbf{W}. \tag{4}$$

where $\mathbf{U} = (U_1, U_2, U_3, U_4, \cdots)$ and $\mathbf{W} = (Q_{W1}, Q_{W2}, Q_{W3}, Q_{W4}, \cdots)$.
Then the matrix $\mathbf{C}$ can be calculated from the following equation:

$$\mathbf{C} = \mathbf{U} \cdot \mathbf{W}^T \cdot (\mathbf{W} \cdot \mathbf{W}^T)^{-1}. \tag{5}$$

In the case with four landmarks, the equation is simply,

$$\mathbf{C} = \mathbf{U} \cdot \mathbf{W}^{-1}. \tag{6}$$

Since the matrix $\mathbf{W}^{-1}$ (or $\mathbf{W} \cdot (\mathbf{W} \cdot \mathbf{W}^T)^{-1}$) is obtained from known
world coordinates of the landmarks, it can be calculated beforehand. Thus,
the problem of estimating the camera parameter $\mathbf{C}$ becomes the problem
of obtaining the matrix $\mathbf{U}$, that is, the image coordinates of four (or more)
landmarks and their depth values $Z_{Ci}$.

Here, the main idea of our registration algorithm is to derive the depth
information of detected landmarks from the rough information of the camera position and orientation obtained from the head tracker. By using viewing matrix $\hat{\mathbf{M}}$ (expressed as a $4 \times 4$ matrix in order to translate from the
world-coordinate system to the camera coordinate system) obtained from
the head tracker, we can estimate the landmark $\hat{Q}_{Ci} = (\hat{X}_{Ci}, \hat{Y}_{Ci}, \hat{Z}_{Ci}, 1)^T$
in the camera-coordinate system. We use the element $\hat{Z}_{Ci}$ as the depth
information of the landmark $Q_i$ in order to get the matrix $\mathbf{U}$.

## (2) With three landmarks

It is possible to obtain the camera parameter $\mathbf{C}$ by using only three landmarks, $Q_1$, $Q_2$, $Q_3$. Let us assume that all $Z$-values of landmarks are
zero in the world-coordinate system, and the coordinate is expressed as
$Q'_{Wi} = (X_{Wi}, Y_{Wi}, 1)^T$. Then the projection of landmarks can be expressed
by the $3 \times 3$ matrix $\mathbf{C}'$ as in the following equation.

$$U_i = \mathbf{C}' \cdot Q'_{Wi} \tag{7}$$

Here matrix $\mathbf{C}'$ is the subset of the matrix $\mathbf{C}$ which omits its third column
(factors related to the $Z$-coordinate), and it is known that the matrix $\mathbf{C}$
can be derived from the matrix $\mathbf{C}'$ [Nakazawa 97].

If three landmarks are detected, the matrix $\mathbf{C}'$ can be obtained from the equation below.

$$\mathbf{C}' = \mathbf{U}' \cdot \mathbf{W}'^{-1} \qquad (8)$$

where $\mathbf{U}' = (U_1, U_2, U_3)$ and $\mathbf{W}' = (Q'_{W1}, Q'_{W2}, Q'_{W3})$. Thus, the problem of estimating the camera parameter $\mathbf{C}$ becomes the problem of obtaining the matrix $\mathbf{U}'$, that is, the image coordinates of three landmarks and their depth values. Just as in the case of four landmarks, the depth values may be obtained from the head tracker.

Note that the landmarks are not required to be placed on an actual $Z = 0$ plane, because the translation matrix $\mathbf{N}$ $(4 \times 4)$ from the plane including the landmarks to the $Z = 0$ plane always exists.

This algorithm compensates for the error of the camera parameter $\hat{\mathbf{C}}$ obtained from the head tracker in order to eliminate the positional errors on the image for the three detected landmarks.

## (3) With two landmarks

In case two landmarks, $Q_1$ and $Q_2$, are detected, the camera parameter $\mathbf{C}$ can be estimated in the same manner, if we set an imaginary landmark $Q_3$.

Suppose that the third (imaginary) landmark $Q_3$ lies at the world coordinate $Q_{W3}$, not on the line $\overline{Q_{W1}Q_{W2}}$. The depth values of landmarks $Q_1$, $Q_2$, and $Q_3$ are obtained from the head tracker. The image coordinate $\hat{q}_3 = (\hat{x}_3, \hat{y}_3)$ of the imaginary landmark $Q_3$ is also calculated from the output of the head tracker. We can define the matrix $\mathbf{U}'$ by using these values.

The matrix $\mathbf{C}$ is compensated to eliminate positional errors for the two detected landmarks using the camera parameter obtained from the head tracker.

## (4) With one landmark

It is possible to compensate for the positional error on a single landmark by setting two imaginary landmarks in the same manner as we did for two landmarks. Note, however, that if only one landmark is detected, the compensation for the positional error on a landmark is accomplished by the method described in Section 4.1.

## (5) Other cases

Many methods other than those described above are applicable to this framework, since it is only necessary to get the depth information of the detected landmarks. For example, if a pair of stereo cameras is mounted on the HMD, the depth information is obtained by evaluating the correspondence between the landmarks detected on the stereo images. If the 3D head tracker is available in addition to the stereo cameras, the camera parameter (that is, registration) is obtained with at least one fiducial in this framework.

## 5   Discussions and Future Studies

Two choices are available for the augmentation of a physical world with a virtual one: video-based and optical-based. A video see-through HMD works using a closed-view HMD and one or two cameras attached to it. The video from the cameras is combined with computer-generated images and displayed on the HMD. This configuration is often used in applications where accurate registration is necessary. This is because 1) digitized video images are available for additional registration methods, and 2) delay, brightness, and contrast between the two spaces can easily be matched in the video see-through [Azuma 97].

On the other hand, an optical see-through HMD uses optical combiners so that users can see the physical world through glasses and simultaneously look at an image displayed on the HMD monitor. Since the physical world is seen directly, there is no time delay in seeing it. In addition, the resolving power of the physical space is limited only by the resolution of the human fovea. This type of system, however, cannot avoid the time lag between the physical world and the virtual one.

In collaborative AR, interactions are not limited to physical-to-virtual or vice versa. Physical-to-physical interaction between participants is also important. In this case, scenes of physical space should be synchronized for all the participants, and the time delays should be minimized in order to match the kinesthetic and visual systems. Thus, we believe that the optical see-through HMD is more suitable for $AR^2$Hockey than the video see-through one.

In order to provide a seamless feeling in the visual simulation process of the MR Living Room, the differences between the image qualities of the physical space and the virtual space should be minimized. For that purpose, the video see-through would be better than the optical see-through. Thus we have adopted a system configuration in which both see-through methods can be used at the same time in order to examine optimal displaying and registration methods.

Many other problems must be solved if we are to make AR systems generic and practical. As for the system configuration of AR, as you may see in Figure 2 and Figure 6, the two systems share common processes. These are the head-tracking process, the landmark-tracking process, the registration process, and the rendering process. They are core of any AR system, regardless of the application. The other processes, including the main process, should be rebuilt according to the application requirements. This is the merit of the distributed AR system configuration we chose for the case studies. We must still confirm, through many more case studies, that this core system actually works.

Registration is another problem. Sensor fusion with physical 3D sensors and tracking landmarks on the captured images is an optimal solution for smart registration. Our registration framework described in Section 4.2 can handle various situations uniformly. More robust algorithms, however, are necessary for practical applications in which the registration area is relatively wide. In addition, prediction of head movement might be effective in decreasing the time lag. Smart researchers in computer vision and sensory technology should be aware that AR is a treasure house of problems demanding creative use of their knowledge.

The study of human factors involved in AR systems is another important subject. The new version of the AR$^2$Hockey system, which has slightly different system configurations, was installed at ACM SIGGRAPH '98. More than 2,000 people played during the conference, and most participants reported that they played the game in much the same way as they did air hockey. During the installation, we changed the augmentation methods and collected a survey from the participants, as well as their scores and playing times. The results of this analysis will be reported at another time.

## Acknowledgments

The authors would like to thank the people of Mixed Reality Systems Laboratory Inc. for their cooperation and useful discussions.

## References

[Azuma 97]   R. Azuma, "A Survey of Augmented Reality," *Presence*, Vol. 6, No. 4, ed. W. Barfield, S. Feiner, T. Furness III, and M. Hirose, MIT Press, Cambridge, MA, pp. 355–385, 1997.

[Bajura 95]   M. Bajura and U. Neumann, "Dynamic Registration Correction in Video-based Augmented Reality Systems," *Computer Graphics and Applications*, Vol. 5, No. 15, IEEE Compter Society Press, Los Alamitos, CA, pp. 52–60, 1995.

[Billinghurst 97]  M. Billinghurst et al., "Shared Space: Collaborative Information Spaces," *Advances in Human Factors/ergonomics, Vol.
21A, Proceedings of HCI International '97*, ed. G. Salvendy, M.
Smith, and R. Koubek, Elsevier, Amsterdam, pp. 7–10, 1997.

[Caudell 92]       T. P. Caudell and D. W. Mizell, "Augmented Reality, an Application of Heads-up Display Technology to Manual Manufactur-
ing Processes," *Proceedings of the Hawaii International Conference on System Sciences*, IEEE Press, Piscataway, NJ, pp.
659–669, 1992.

[Feiner 93]        S. Feiner et al., "Knowledge-based Augmented Reality," *Communications of the ACM*, Vol. 36, No. 7, New York, NY, pp.
52–62, 1993.

[Hoshi 96]         H. Hoshi et al., "Off-axial HMD Optical System Consisting of
Aspherical Surfaces without Rotational Symmetry," *Proceed-
ings of SPIE*, Vol. 2653, Bellingham, WA, pp. 234–242, 1996.

[Mellor 97]        J. P. Mellor, "Realtime Camera Calibration for Enhanced Reality Visualization," *Proceedings of CVRMed '95*, ed. N. Ayache,
Springer-Verlag, Berlin, pp. 471–475, 1995.

[Milgram 94]       P. Milgram and F. Kishino, "A Taxonomy of Mixed Reality Visual Display," *Institute of Electronics, Information and Communication Engineers, Transactions on Information & Systems*,
Vol. E77-D, No. 12, Tokyo, pp. 1321–1329, 1994.

[Nakazawa 97]      Y. Nakazawa et al., "A System for Composition of Real Moving
Images and CG Images Based on Image Feature Points," *the
Journal of the Institute of Image Information and Television
Engineers*, Vol. 51, No. 7, ed. M. Machida, Tokyo, pp. 1086–
1095, 1997, (In Japanese).

[Ohshima 98]       T. Ohshima et al., "AR$^2$Hockey: A Case Study of Collaborative Augmented Reality," *Proceedings of VRAIS '98*, IEEE
Computer Society Press, Los Alamitos, CA, pp. 268–275, 1998.

[Rekimoto 96]      J. Rekimoto, "TransVision: A Hand-held Augmented Reality
System for Collaborative Design," *Proceedings of VSMM '96*,
Gifu, Japan, pp. 85–90, 1996.

# Computer-vision-enabled Ophthalmic Augmented Reality: A PC-based Prototype

Jeffrey W. Berger and David S. Shin

## 1 Introduction

Although there has been an explosive increase in the development of virtual reality (VR) applications in medicine, augmented reality (AR) paradigms may offer far greater utility, but have received much less attention [Tang 98] [O'Toole 95]. Whereas VR is the construction of a synthetic environment, AR extracts information from the real world and augments it.

An early study described the superposition of ultrasound images on the abdomen, using a position-tracked, see-through, head-mounted display [Bajura 92]. Recently, there has been great interest in neurosurgical applications of AR. Specifically, the registration of CT, MR, and PET images on a living patient in the operating room may greatly facilitate surgical planning and execution [Gleason 94] [Edwards 95] [Grimson 96]. Typically, registration of the radiographic data with the patient is accomplished by tracking fiducial markers placed on the skin surface, or through other complex tracking strategies.

We have initiated investigations toward the design and implementation of an ophthalmic augmented reality environment for two reasons. First, to permit overlay of previously stored photographic and angiographic images onto the real-time biomicroscopic slitlamp fundus (retinal) image in order to guide treatment for eye disease — for example, to define better and visualize the edges of choroidal neovascularization (CNV) in age-related macular degeneration (AMD) — and to facilitate real-time image measurement and comparison. Second, to take advantage of the fact that the fundus biomicroscopic image is quasi-two-dimensional: computational demands for computer-vision-enabled image overlay in two dimensions are far less intense than in three-dimensional (for example, neurosurgical) applications. Thus, this simpler paradigm may be used as a model system for more complex computer-vision-enabled augmented reality environments, particularly those in the medical field [Sharma 95] [Uenohara 95].

Diabetic macular edema and AMD are the two major causes of visual loss in developed countries. While laser therapy for these and other diseases

19

has prevented loss of visual function in many individuals, disease progression and visual loss following suboptimal treatment is common. For AMD, there is unambiguous evidence that incomplete laser photocoagulation of the border of CNV is associated with an increased risk of further visual loss, while treatment beyond the borders unnecessarily destroys viable, central photoreceptors, further degrading visual function [Berger 99]. Building on the recommendations proposed in the Macular Photocoagulation Studies [Berger 98], clinicians generally attempt to correlate angiographic data with biomicroscopic images using crude, time-consuming, potentially error-prone methods.

In a recent practical review [Neely 96], the author describes such a method: "to assist you in treatment (of neovascular AMD), project an early frame of the fluorescein angiogram onto a viewing screen (next to the patient). Use the retinal vessels overlying the CNV lesion as landmarks. I suggest tracing an image of the CNV lesion and overlying vessels onto a sheet of onion skin paper. It takes a little extra time, but I find it helps to clarify the treatment area."

Identifying precisely the treatment border during laser therapy by correlating the biomicroscopic image with fluorescein angiographic data (where the lesion extent is better delineated, see for example Figure 1) should be beneficial for maximizing post-treatment visual function. Diagnosis and

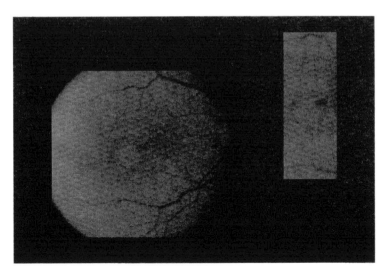

**Figure 1.** Left, thirty-degree fundus photograph. Blood vessels emanate from the optic nerve partially seen at the right edge of the image. Right, simulated slitlamp image of the same eye.

treatment relies on synthesizing clinical data derived from fundus biomicroscopy with angiographic data, but methods for correlating these data, and for direct guidance of laser therapy, are not well-developed.

We have previously presented our early work relevant to design considerations for a computer-vision-enabled, fundus-landmark-driven, ophthalmic augmented reality environment [Berger 97]. We have since migrated from a Unix platform to a Windows NT/Personal Computer (Intel) platform, and have modified our algorithms considerably. Moreover, our initial investigations explored ophthalmic image overlay in an operating microscope environment, whereas we are now constructing a working prototype on a slit-lamp biomicroscope platform; the slitlamp is the standard instrument for ophthalmic diagnosis and treatment. In this report, we describe the design and construction of a working slitlamp-based prototype and report on initial experience in model eyes.

## 2 Methods

While some augmented reality environments require fiducial markers for tracking the real-world object of interest, the eye fundus is ideal for computer-vision-enabled image overlay in that retinal blood vessels are omnipresent, high-contrast, sharply-edged features highly suitable for tracking. The geometry of the blood vessels is identical in the real-time image and in the previously-stored photographic and angiographic images to be overlayed on the real-time image.

However, design and implementation of the system described requires a solution to a number of challenging problems. First, the system must be robust and tolerate a small area of fundus illumination. While stored fundus images generally subtend at least thirty degrees of retina, slitlamp examination illuminates only a small, usually rectangular portion of the retina (Figure 1), albeit with portions of blood vessels present. Moreover, the system must respond rapidly to moderate changes in eye position during evaluation and treatment, and the computer rendering must be fast, and ergonomically well-tolerated. Welch and coworkers [Markow 93] [Barrett 94] [Barrett 96] [Becker 95] describe algorithms for real-time, retinal tracking.

However, their tracking algorithms require multiple templates to follow retinal vessels at several fundus locations, and therefore require a large illumination area — tracked vessels must be visible at all times. This requirement is not consistent with our goal of allowing for narrow beam illumination, and is more suitable for a fundus camera with monocular, wide-angle viewing. Further, their tracking algorithms limit the search to a small area surrounding the previously tracked position and are not tolerant of large changes in fundus position, as might be encountered during a slitlamp fundus, biomicroscopic examination.

## 3  Hardware

Our hardware configuration is depicted in Figure 2. The slitlamp biomicroscope (Nikon NS-1V) permits binocular, stereoscopic examination of one eye of a patient. The image of one of the oculars is split to a CCD camera (Optronics LE-470), and sent to a framegrabber (Matrox Corona) and personal computer (Dell Dimension, 333-MHz Pentium II Processor with MMX, running Windows NT 4.0).

Although our early studies were designed around a Sun workstation, the current implementation is built around an IBM-compatible personal computer to permit rapid incorporation into the clinical environment. Temporal considerations dictate speed requirements. For effective image overlay with the potential for image-guided laser therapy, update times of approximately 100 ms are required. This can be accomplished with video rate acquisition (30 frames per second, 33 ms), with 67 ms available for image processing and rendering.

The development of PC-compatible framegrabber boards with on-board processing including edge-detection, convolution, and correlation permits

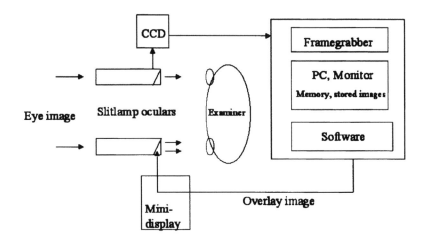

**Figure 2.** Hardware schematic.

rapid processing. However, MMX-optimized software, for example, Matrox Imaging Library (MIL 5.0, Matrox Inc., Quebec, Canada), allows for efficient computation and may prove sufficient for this application, obviating the need for more expensive and less versatile hardware-based processing.

Accordingly, our platform is designed around a personal computer and a framegrabber board with memory but without on-board processing (currently Matrox Corona), driven by MIL software. Based on the real-time acquired image, the stored images may be rendered (as selected by the examiner) on a miniature display registered with the real-time biomicroscopic fundus image, and presented through the second ocular of the biomicroscope. Although it is useful to track and render the real-time image superimposed on previously stored photographic and angiographic images on a computer monitor, image overlay directly onto the real-time image in the slitlamp biomicroscope environment allows for a very natural augmentation of the standard platform for ophthalmic diagnosis and treatment.

We have been investigating the modification of our custom-made beamsplitter to permit the incorporation of a miniature CRT or LCD display. Our initial studies suggest that a one-inch diagonal, VGA-resolution ($640 \times 480$) LCD display is of sufficient resolution and brightness for high-quality image overlay. Eventually, a laser-capable slitlamp will serve as the platform, permitting image-guided laser therapy [Shin 99].

# 4   Software

Efficient computation is achieved by dividing the software requirements into real-time and non-real-time components (Figures 3 and 4). The objective is to limit the computation performed during real-time function. In non-real-time, angiographic and fundus photographic images are mutually registered with high fidelity. Since photographic and angiographic data may vary considerably in intensity — e.g., the blood vessels are dark in a monochromatic image and in the early angiographic phases but bright in the late angiographic phase — intensity-based registration algorithms are unsuitable. Edge-based registration algorithms permit successful image registration of color, monochromatic, and angiographic data, following image pre-processing with smoothing, edge detection [Canny 86], and thresholding. We have previously demonstrated the applicability of Hausdorff distance-based algorithms [Berger 97], and have similarly successfully applied elastic matching techniques [Bajcsy 89], as well as simple, rigid body manipulations (rotation, translation, and scale) for mutual registration.

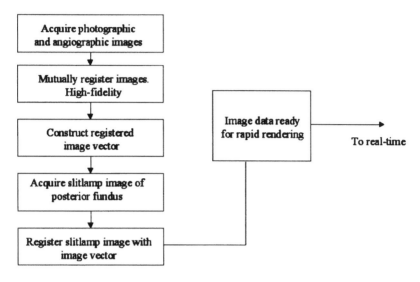

**Figure 3.** Software schematic, non-real time component.

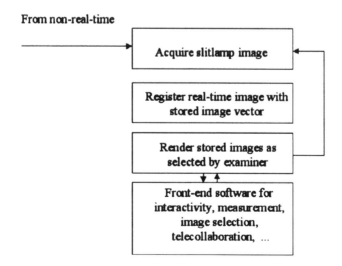

**Figure 4.** Software schematic, real-time component.

For robust performance while tolerating eye movement, a map of the posterior eye fundus is created by montaging partially overlapping fundus images. Photo montages may be created with a modified, bidirectional, partial-Hausdorff, distance-based algorithm [Huttenlocher 93] [Rucklidge 95] [Berger 97]. Since spurious edges may corrupt edge-based tracking functions, the montage is smoothed with a convex combination that puts greater weight on pixels closer to the center of an image, resulting in a seamless photo montage.

For real-time registration, we are now capitalizing on Matrox Imaging Library functionality in a Visual C++ environment. Benchmarks for performance on a Pentium II, 400-MHz processor with MMX technology (the current, commercially available, state-of the-art, single-processor desktop system) are 9.6 ms for Sobel edge detection on a $512 \times 512 \times 8$ bit image, and 17.3 ms for normalized gray-scale correlation for finding a $32 \times 32$ model in a $512 \times 512$ image with sub-pixel accuracy. By combining native MIL functions with custom-developed C++ routines, we are able to exploit the speed and ease of MIL and the pluripotentiality of C++.

In real time, the acquired slitlamp fundus image is then mapped to the stored photo montage, allowing for registration with fast template matching and rendering. We are currently using native MIL functions which allow for a multiresolution, normalized gray-scale, template match searching over translation and rotation. With the knowledge that there is limited rotational misalignment between the real-time and stored images, we restrict the angular search to +/- 8 degrees. For further efficiency, and to allow real-time performance, we interactively adjust scale prior to tracking. Search over scale is straightforwardly incorporated at the expense of increased search time.

## 5 Results

Each element of the system is operational, and the system is being further developed and refined. At present, and pending final construction of the beamsplitter allowing for direct image overlay, development is proceeding with image rendering on a computer monitor. We have developed a graphical user interface for rendering of the real-time video image as well as previously stored images.

We have demonstrated robust, near-video-rate tracking in studies with a model eye. Images of the model eye may be montaged, or a wide-field view may be acquired. With search for translation and rotation, tracking rates of 5 to 10 Hz are achieved, with misregistration on the order of one to five pixels in most studies (Figures 5 and 6).

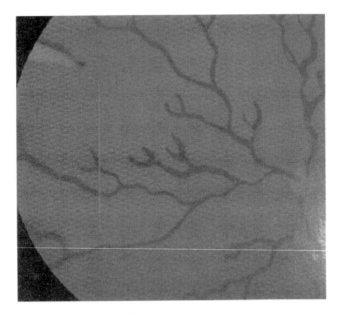

**Figure 5.** 60-degree image of model eye.

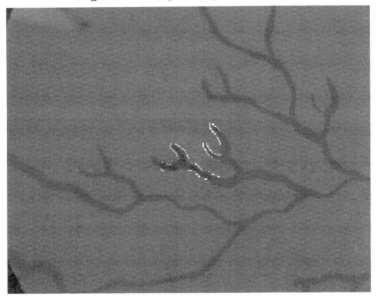

**Figure 6.** Slit image (central) tracked on 60-degree image, with the remainder rendered from memory registered with the illuminated slit image.

# 6   Discussion

This paper describes progress towards the construction and implementation of a computer-vision-enabled, slitlamp-based, ophthalmic augmented reality environment. As an analog to the operating microscope, the slitlamp biomicroscope is perfectly suited to serve as an imaging and display conduit for AR applications. Whereas many AR systems require a dedicated graphical apparatus (for example, a heads-up display), the routine environment for diagnosis and treatment of eye disease is a binocular platform straightforwardly modified with beamsplitters.

Three-dimensional augmented reality applications (for example, obstetrical or neurosurgical applications) require 3D–3D registration of graphical information with real-world coordinates. Our application is quasi-two-dimensional, so that computational demands are far less stringent than in three-dimensional image registration tasks, potentially allowing for efficient real-time performance. However, while neurosurgical patients do not ordinarily move during operative procedures, the eyeball is usually somewhat kinetic during ophthalmic diagnosis and treatment; with few exceptions, ophthalmic examination and laser treatment are performed with topical anesthesia, while the patient holds his or her eye as still as possible. Therefore, ophthalmic AR requires real-time tracking.

In a previous paper, we described development of a prototype built around a Sun workstation and on an operating microscope platform [Berger 97]. In this report, we extend our previous work to a PC-based system, in the preferred implementation of a slitlamp biomicroscope-based platform. While we previously used the Hausdorff distance as the similarity metric facilitating both montaging and real-time tracking, we now use the Hausdorff distance for montaging only, and track in real-time using modified Matrox Imaging Library functions. Hausdorff distance-based methods permitted tracking at 0.5 Hz, which was not sufficient for clinical utility. However, our current implementation permits 5 to 10-Hz performance, conferring clinical utility.

Incorporation of our beamsplitter with miniature display, fully designed and now under construction, will allow for a natural augmentation of the standard platform for ophthalmic diagnosis and treatment. We then expect to validate this method for improved diagnosis and treatment in the clinical environment. In addition, we are developing a platform for ophthalmic telemedicine and telecollaboration [Hariprasad 99] which is built on and intimately coupled with the AR platform described here, and we expect to implement this platform in the near future.

# Acknowledgments

Supported by Career Development Award from Research to Prevent Blindness (JWB) and NIH K-08 00374 (JWB).

# References

[Azuma 93]        R. Azuma. "Tracking Requirements for Augmented Real-
                  ity," *Communications of the ACM*, Vol. 36, No. 7, pp. 50–
                  51, 1993.

[Bajcsy 89]       R. Bajcsy and S. Kovacic. "Multiresolution Elastic Match-
                  ing," *Computer Vision, Graphics and Image Processing*, Vol.
                  46, pp. 1–21, 1989.

[Bajura 92]       M. Bajura, H. Fuchs, and R. Ohbuchi. "Merging Virtual
                  Objects With the Real World: Seeing Ultrasound Imagery
                  Within the Patient," *Computer Graphics*, Vol. 26, pp. 203–
                  210, 1992.

[Barrett 94]      S.F. Barrett, M.R. Jerath, H.G. Rylander, and A.J. Welch.
                  "Digital Tracking and Control of Retinal Images," *Optical
                  Engineering*, Vol. 33, pp. 150–159, 1994.

[Barrett 96]      S.F. Barrett, C.H.G. Wright, M.R. Jerath, S. Lewis, B.C.
                  Dillard, H.G. Rylander, and A.J. Welch. "Computer-aided
                  Retinal Photocoagulation System," *Journal Biomedical Op-
                  tics*, Vol. 1, pp. 83–91, 1996.

[Becker 95]       D.E. Becker, J.N. Turner, H. Tannenbaum, and B. Roysam.
                  "Real-time Image Processing Algorithms for an Automated
                  Retinal Laser Surgery System," *Proceedings IEEE 2nd Inter-
                  national Conference on Image Processing*, pp. 426–429, 1995.

[Berger 97]       J.W. Berger, M.E. Leventon, N. Hata, W. Wells, and R.
                  Kikinis. "Design Considerations for a Computer-vision En-
                  abled Ophthalmic Augmented Reality Environment," *Lec-
                  ture Notes in Computer Science. CVRMed-MRCAS 97*,
                  ed. J. Troccaz, E Grimson, R. Mosges, Vol. 1205, pp. 399–
                  408, 1997.

[Berger 98]       J.W. Berger, M.G. Maguire, and S.L. Fine. "The Macular
                  Photocoagulation Studies," *Clinical Trials in Ophthalmol-
                  ogy*, ed. P. Kertes, M. Conway, Williams, and Wilkins, Bal-
                  timore, 1998.

[Berger 99]       J.W. Berger and S.L. Fine. "Laser Treatment for Choroidal
                  Neovascularization," *Age-related Macular Degeneration*, ed.
                  J.W. Berger, M.G. Maguire, and S.L. Fine, Mosby, St.
                  Louis, 1999.

[Bowskill 95]        J. Bowskill and J. Downie. "Extending the Capabilities of the Human Visual System. An Introduction to Enhanced Reality," *Computer Graphics*, Vol. 29, pp. 61–65, 1995.

[Canny 86]           J.F. Canny. "A Computational Approach to Edge Detection," *IEEE Transactions Pattern Anallysis and Machine Intelligence* Vol. 8, pp. 34–43, 1986.

[Caudell 94]         T.P. Caudell. "Introduction to Augmented and Virtual Reality," *Proceedings SPIE (Telemanipulator and Telepresence Technologies)*. Vol. 2351, pp. 272–281, 1994.

[Edwards 95]         P.J. Edwards, D.L.G. Hill, D.J. Hawkes, R. Spink, A.C.F. Colchester, A. Strong, and M. Gleeson. "Neurosurgical Guidance Using the Stereo Microscope," *Proceedings First International Conference on Computer Vision, Virtual Reality and Robotics in Medicine*, Nice, pp. 555–564, 1995.

[Feiner 93]          S. Feiner, M. Macintyre, and D. Seligmann. "Knowledge Based Augmented Reality," *Communications of the ACM*, Vol. 36, pp. 53–61, 1993.

[Gleason 94]         P.L. Gleason, R. Kikinis, D. Altobelli, W. Wells, E. Alexander, P. Black, and F. Jolesz. "Video Registration Virtual Reality for Nonlinkage Stereotactic Surgery," *Sterotactic Functional Neurosurgery*, Vol. 63, pp. 139–143, 1994.

[Grimson 96]         W.E.L. Grimson, G.J. Ettinger, S.J. White, T. Lozano-Prez, W.M. Wells III, and R. Kikinis. "An Automatic Registration Method for Frameless Stereotaxy, Image Guided Surgery, and Enhanced Reality Visualization," *IEEE Transactions Medical Imaging*, Vol. 15, pp. 129, 1996.

[Hariprasad 99]      R. Hariprasad, D. Shin, and J.W. Berger. "An Intelligent Platform for Ophthalmic Teaching, Telemedicine, and Telecollaboration," *Proceedings Medicine Meets Virtual Reality 7*, IOS Press, Amsterdam, 1999.

[Huttenlocher 93]    D.P. Huttenlocher, G.A. Klanderman, and W.J. Rucklidge. "Comparing Images Using the Hausdorff Distance," *IEEE Transactions Pattern Analysis Machine Intelligence*, Vol. 15, pp. 850–863, 1993.

[Markow 93]          M.S. Markow, H.G. Rylander, and A.J. Welch. "Realtime Algorithm for Retinal Tracking," *IEEE Transactions Biomedical Engineering*, Vol. 40, pp. 1269–1281, 1993.

[Milgram 94]         P. Milgram, et al. "Augmented Reality: A Class of Displays on the Reality-Virtuality Continuum," *Proceedings SPIE: Telemanipulator and Telepresence Technologies*, Vol. 2351, pp. 282–292, 1994.

[Neely 96]          K.A. Neely. "How to be More Successful in Laser Photoco-
                    agulation," *Ophthalmology Times*, pp. 103–108, June 1996.

[O'Toole 95]        R.V. O'Toole, M.K. Blackwell, F.M. Morgan, L. Gregor,
                    D. Shefman, B. Jaramaz, DiGioia, and T. Kanade. "Image
                    Overlay for Surgical Enhancement and Telemedicine," *In-
                    teractive Technology and the New Paradigm for Healthcare*,
                    ed. K. Morgan et al. IOS Press, Amsterdam, 1995.

[Rucklidge 95]      W.J. Rucklidge. "Locating Objects Using the Hausdorff
                    Distance," *Proceedings IEEE International Conference on
                    Computer Vision*, pp. 457–464, 1995.

[Sharma 95]         R. Sharma and J. Molineros. "Role of Computer Vision in
                    Augmented Virtual Reality," *Proceedings SPIE*, Vol. 2409,
                    pp. 220–231, 1995.

[Shin 99]           D.S. Shin and J.W. Berger. "Image Guided Macular Laser
                    Therapy," *Proceedings SPIE*, in press.

[Tang 98]           S.L. Tang, et al. "Augmented Reality Systems for Medical
                    Applications. Improving Surgical Procedures by Enhancing
                    the Surgeon's 'View' of the Patient," *IEEE Engineering in
                    Medicine and Biology*, pp. 49–58, May/June 1998.

[Uenohara 95]       M. Uenohara and T. Kanade. "Vision-based Object Regis-
                    tration for Real-time Image Overlay," *Proceedings 1st Con-
                    ference on Computer Vision, Virtual Reality and Robotics
                    in Medicine*, Nice, pp. 13–22, 1995.

# Augmented Reality for Construction Tasks: Doorlock Assembly

Dirk Reiners, Didier Stricker, Gudrun Klinker, and Stefan Müller

**Abstract.** *Augmented reality (AR) is a technology that integrates pictures of virtual objects into images of the real world. This technology will be useful to industry when the technical problems have been solved, and when the benefits clearly outweigh the work required to use it. Furthermore, researchers must answer the important question of how AR might be integrated into the information technology infrastructure of a company. This paper describes an augmented reality demonstrator for the task of doorlock assembly in a car door. The demonstrator was developed as a practical, realistic application that could convey to a casual observer the concepts behind AR. A new, fast, and robust optical tracking algorithm was developed and integrated into a three-dimensional animation and rendering system, thereby creating a real-time fully three-dimensional HMD-based training application that teaches users how to assemble the doorlock into the door. The system was demonstrated to the general public at the Hannover Industrial Fair 1998; this demonstration of AR to a large, non-expert audience generated greater interest in this new area. There are still some technological problems to solve, but in order to get industrial partners interested in investing in this technology, researchers must make clear its possible benefits and its ability to be integrated into the whole company.*

## 1 Introduction

Augmented reality is a technology that integrates pictures of virtual objects into images of the real world. These images can be taken from a camera, or the user's direct view on the world can be augmented with the use of a see-through, head-mounted display. There are still some technological problems to solve, but in order to get industrial partners interested in investing in this technology, researchers must make clear possible benefits and ability to be integrated into the whole company.

Many companies are trying to move larger portions of their design and construction processes into the computer. Physical prototypes are being

31

replaced by virtual prototypes for packaging, assembly, and security evaluations [Dai 96]. This is especially evident in the airplane and car construction companies, where prototypes are costly and time to market is a significant factor. A side-effect is that a large number of three-dimensional models become available at little or no cost.

AR can use this growing quantity of digital product information to help people who work in a world where new products are introduced more and more rapidly. One key consequence of reduced time to market, shorter product cycles, and smaller production runs is an increased demand for training. Assembling a complex machine like a car or a plane is a difficult process that takes time to learn. If that training is only useful for a small number of production runs, the efficiency of the worker decreases. The situation is even worse for service personnel, since faster product cycles don't necessarily go along with shorter life cycles. Thus, a service person is confronted with a rapidly growing number of generations of systems and must be able to handle all of them. Given the widespread distribution of service personnel for such items as cars, centralized training has become difficult, expensive, or, in some cases, impossible.

The availability of three-dimensional models allows us to train and instruct personnel in novel ways that are more direct and more intuitive than written instructions or even electronic hypertext manuals. Three-dimensional animated instructions can be integrated into the surrounding environment at the exact place where the action has to be performed, so that no mental transfer is needed, and the action can be observed from different viewpoints in order to help the user understand the spatial relationships involved.

This paper describes an implementation of an augmented reality system that builds this connection between virtual prototyping and real assembly tasks — in this case, the task of assembling a doorlock into a car door. In Section 2, we review previous work on the use of AR in assembly and maintenance tasks and in Section 3 we introduce the application task. Following that is a description of the required hardware (Section 4.1) and software (Sections 4.3 and 4.4) systems, together with motivations and alternatives. We have developed a new optical tracking system that is described in Section 4.2. The system was shown to the general public at the Hannover Industrial Fair 1998 (Section 5); as a conclusion, experiences and open areas for further work are analyzed in Section 6.

## 2   Previous Work

For quite some time, assembly and maintenance tasks have been a target area of AR demonstrations. One of the first demonstrations used augmented reality for a photocopier maintenance task [Feiner 93]. This was done using wireframe graphics and a monochrome, monoscopic HMD. Head and object tracking were achieved using ultrasonic trackers. The graphics used were rather simple and were not taken from any CAD system, since the main objective was to extend an existing two-dimensional, automated instruction generation system to an augmented environment. This scenario was recently revived by Mann [Mann 98], this time tracked optically.

Another application in the construction area was the one presented by Webster [Webster 96], in which AR technology was used for spaceframe assembly support. Tracking was done using an active optical system with blinking LEDs, coupled to a passive inertial system. The graphics used were rather stylized, since the objective was not to convey a sense of spatial presence, but to convey simple instructions.

Boeing has used an AR system for wire bundle assembly support. The tracking was done using an inertial/ultrasonic system. The augmentations (2D lines showing the path of the cable to be added to the bundle) were simple and could be generated from a database without manual involvement or a CAD system [Caudell 92].

## 3  Application

The application targeted in this demonstration is the assembly of the door-lock in a car door (see Figure 2). The virtual environment group of our department realized this task would set a precedent for virtual prototyping and assembly feasibility for BMW (see Figure 1). Thus, CAD data taken from the system used to construct the actual objects was available not only for the lock but also for the whole door.

The task of assembling a doorlock is a challenge that requires significant planning and dexterity. It is spatial and three-dimensional in nature, and

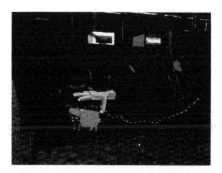

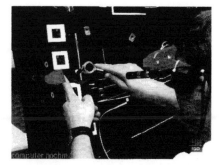

**Figure 1.** Virtual doorlock assembly.     **Figure 2.** Two-handed screw fixing.

the movement of the hand holding the lock in the small space inside the door necessitates precise preparation and movement. A little lever which is not easily visible from the outside (see Figure 10) has to engage in the lock deep inside the door, and must be pushed to the right position beforehand — otherwise, the final engagement of the lock will be impossible. Furthermore, the space inside the door is just big enough for the lock. Thus, the lock has to be gripped in such a way that the hand does not get stuck halfway through (see Figure 11). This tight space gives strong haptic feedback which facilitates the task, but it demands a good three-dimensional understanding of the space.

Once the lock's antenna rests in its final position, three screws must be tightened to lock it into place (see Figure 2). The tasks of inserting the antenna into the door compartment and of tightening the screws demand the use of both hands. Thus a hands-free interaction method is needed in order to control the instruction flow.

## 4   System

### 4.1   Hardware

The system was built using common, off-the-shelf hardware: a standard SGI O2 with a 180-MHz R5k processor and 128 MB of memory. The machine has very good video capabilities and provides reasonably fast rendering, although it is not comparable to higher-end SGI machines used for virtual reality applications like the one mentioned in Section 3.

The strong spatial nature of the task demanded a display that could convey spatial information, an HMD being the logical choice. The HMD we used is a Virtual I/O i-glasses set with an attached Toshiba IK-M48PK camera using 7.5-mm optics, as shown in Figure 3. The headset is a standard and affordable piece of equipment that allows see-through and feed-

**Figure 3.** Head-mounted display with camera.

through use, making it useful for a variety of applications. The camera is reasonably small and light enough to be worn on the head without undue strain for the user, while still giving very good quality for a single-CCD camera.

The two-handed action needed for the task required a hands-free interaction technique. Thus, a voice-command-driven interface was used. It runs on a separate machine — a standard laptop running Windows 95 and IBM Voice-Type-based speech recognition software. It is connected to the O2 via RS-232, which is adequate for the transmission of the short recognizable commands.

## 4.2  Tracking

Tracking the user's actions has been a major focus of the AR community as well as of the virtual reality community. Different technical solutions to the problem of measuring the user's motion have been developed.

The magnetic trackers, which are very often found in VR applications, were not considered applicable for the task at hand. This is due to their sensitivity to metallic objects in the surrounding area, e.g., the door. Given that the final deployment location for the application would be a construction hall, where everything is metallic, an alternative solution had to be found.

An optical method was considered superior. There has been previous work on tracking using optical markers. Neumann et al. as well as State et al. use color-coded dots [Neumann 96]; if a larger number of markers is needed, rings are used [Neumann 97] [State 96]. Starner et al. [Starner 97] use a color-coded bit pattern, but they don't do a full camera reconstruction. Their application, using 2D text and graphics annotations, can be achieved using only image position and orientation coupled with size. Vallino et al. [Kutulakos 96] don't need to reconstruct a full camera parameter set from their recognized feature points, since they are using a non-projective rendering scheme. The closest approach to ours is that of [Koller 97], which uses the same markers and has a similar goal.

For the given task, full three-dimensional position and orientation were considered mandatory, so our tracker had to be able to do a full camera calibration.

**Targets**  Our tracker uses square targets (see Figure 4) made from black cardboard. These targets are attached to the door (see Figure 5), as this allows the door to move around freely without requiring a recalibration or remeasurement. The relative position of the targets is measured beforehand, using a tape measure. Their size can vary widely; the ones used on

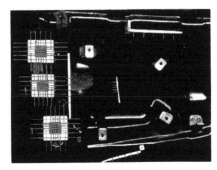

**Figure 4.** Square target.          **Figure 5.** Door with tracked features.

the door were $2.5 \times 2.5$ cm. They are identified by a code of small red squares attached to them at known positions. Two rows of four IDs are possible, allowing 256 different squares to be used. The symmetric nature of the IDs means that some of the combinations are not unique — i.e., in the case of rotations. For the door application and most others this is not a problem, since users rarely rotate their heads more than 45 degrees relative to the horizon. If it does pose a problem, the offending IDs can be left out so that all used IDs are rotationally unique.

**Initialization** At initialization time, the full picture is searched for black squares with valid markers. From the position of their corners in the image, as determined by the intersections of their edges, the camera parameters are calculated using the technique of either Tsai [Tsai 86] or Weng et al. [Weng 92]. Every identified square adds four points to the calibration algorithm. Thus, since a full calibration requires 11 points, three squares are needed. The camera we used (see Section 4.1) has a fixed focus and fixed internal parameters, allowing a variant of Tsai's technique to be chosen that only calculates the external parameters (position and orientation), thus giving better precision and stability, and requiring only two squares for initialization.

**Operation** After this initial step, which takes about 200 ms on our hardware, all further measurements are done incrementally. The 2D position of the squares' edges are extrapolated based on the position and speed in the previous frames; the position and orientation of the real edge is found using a local search; and the new camera parameters are calculated using an error minimization technique (for details see [Stricker 98]). Only a small part of the image has to be examined, so the speed can go up to 25 Hz, which is quite adequate for head-mounted usage.

Using only lines for calibration has two additional benefits. The first is that the red IDs don't have to analyzed anymore, so it is possible to extend the operational range beyond the point where they are no longer discernible. The second is that a more general form of target can be used — any high-contrast line. Given the 3D position and orientation, any line can be used to add information to the tracking process, making it more stable and robust. For this reason, a number of white tape marks have been added to the door (see Figure 5), which are used in conjunction with the white squares surrounding the black squares. The tape marks are significantly less disturbing visually than the rather prominent squares, and they still add useful information.

The IDs are not checked in the incremental mode; theoretically, after a fast move, a tracking edge could get locked onto another, incorrect feature edge in the image. This is detected by measuring the error in the calibration. If the error gets too big (because some lines were lost or locked onto wrong image features), a new initialization step must be performed in order to return to data known to be good.

**See-through Calibration**  One of the major problems for see-through applications is the calibration between the head tracker (in our case the head-mounted camera) and the user's eyes [Janin 93]. The solution we decided upon was to reuse the know-how and algorithms we had developed for the camera calibration problem which we could do by abusing the user as a feature finder. For calibration purposes, the user is asked to click manually (using the mouse) on the corners of the calibration targets as he sees them through his HMD. This provides enough information to calculate a calibration for the user's eye. Given additional constraints valid for the two eyes (that they have the same orientation except for a rotation around the up-axis), it is possible to calculate with appropriate precision the relative position and orientation in relation to the camera. This method turned out to work quite well, but it is dependent on the users' keeping their heads very still during the calibration procedure. To facilitate this, we built a simple headrest, which improved calibration quality considerably.

The variation in parameter sets for different users was small enough for one person's parameters to be usable for another person. For best results, individual calibration is still needed, but for a quick impression it is acceptable simply to exchange the HMD.

## 4.3  Data Acquisition

One of the goals of the project was to see how much of the data required for an augmented reality application could be taken from other sources.

**Figure 6.** Original (12,156 polygons) and reduced lock models (656 polygons)

As it turned out, all the model data could be taken from the virtual reality application that had already been developed (see Section 3). If the AR application were run on a high-end graphics machine like the Reality Engine 2 used for the VR application, the data could have been used directly. The O2 we used has very good video capabilities, but its rendering power is much more limited than that of the Reality Engine. Thus the models had be to reduced in complexity in order to prevent an unacceptable slowdown of the system. The models were reduced semi-automatically to a point where the graphical complexity was acceptable without compromising rendering quality (see Figure 6).

The door model needed for occlusion handling (as shown in Figure 14) was created by deleting unnecessary parts and remodeling slightly to remove polygonally expensive but unused holes.

The animation paths used for the animated instructions were also taken from a performance of the task in the virtual world. Due to inaccuracies in the magnetic tracking system used in the VR application, the virtual data was not of high enough quality to be used directly in an educational demonstration. Instead, some frames were chosen manually and used as keyframes for the animated instructions. This gave a very smooth and continuous animation with very little effort.

## 4.4   Software

The software architecture, as depicted in Figure 7, revolves around the central tracker loop. The tracker deals with reading and analyzing the video image in order to calculate the camera calibration parameters, which are translated into OpenGL set-up matrices. A second, coupled component provides the application-dependent augmentations and user interactions.

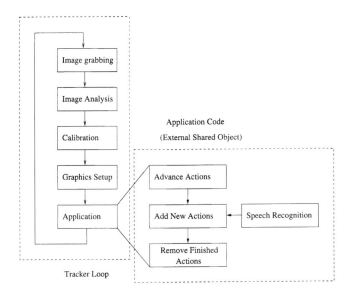

**Figure 7.** Software architecture.

**Distribution and Performance** Experimentally, the system was distributed between two machines: one for tracking (the SGI O2) and one for rendering (an SGI Onyx RE2). The calibration information was sent via Ethernet between the two machines. But the lag time resulting from the system and the network proved too long to allow this approach to be useful for head-mounted applications. Thus, for the final application, a single system as described in Section 4.1 was used. The tracking itself runs at 20-25 Hz for the given set-up. The speed drops to 13-15 Hz due to the additional burden of rendering the augmentations, which, together with a frame of display latency, adds 100-ms end-to-end delay. This is right at the border of acceptability for head-mounted applications, and the system would benefit from a faster update rate; but for demonstration purposes, the performance was acceptable. Given the expected rise in graphics performance, next-generation hardware will be able to run the application comfortably.

**See-through vs. Feed-through** Another advantage of the closely connected system is the possibility of letting the tracker system draw the video frame used for calibration as a background, before passing control to the application. In turn, this allows switching between see-through and feed-through calibration within the same system. Furthermore, the decision of whether to use feed-through or see-through does not need to be made until a later stage in the development cycle, so the decision can be based on tests in real situations. When we used feed-through, the offset between the head-mounted camera and the user's eye, in connection with the very close working range for the task, made the system feel very awkward. The difference between the kinesthetic information coming from the arms and hands and what could seen in the display proved to be very distracting. Another major disadvantage of the feed-through approach was the presence of only one camera, which forced a monoscopic feed-through mode. For the given task, the lack of stereoscopic depth made performing the task very difficult. Thus, see-through was the preferred mode of operation.

**Application Code Integration**   The actual application is attached to the tracker by means of a simple procedural interface. Three functions (init, loop, exit) are needed to attach a program to the AR system. When the loop function is called, the application program can call OpenGL commands in order to generate an appropriate image and react to user input. For our system, OpenInventor was used for rendering. It has a clean and simple interface for rendering to an arbitrary OpenGL window, and, for the simple models employed, its performance deficiencies were acceptable.

This division between the tracker and the application eased application development considerably, due to the possibility of using a simple, simulator program that adhered to the same procedural interface and used prerecorded calibration data. Thus, actions were easily reproducible, and development could be done on any OpenGL-capable machine, independent of any special video hardware.

**Application Code**   The application code as implemented for this scenario uses the action graph depicted in Figure 8. The whole task is divided into separate steps that are short enough to be executed by the user without the need for breaks. The instructions to be given for each step are structured as an acyclic directed graph, where every structural node (oval)

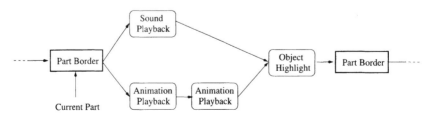

**Figure 8.** Action graph

has an associated type and duration. Branches in the graph are executed in parallel, and joins are synchronization points such that all actions leading into them have to finish before the next action is executed.

Possible actions include sound and animation playback, as well as object highlighting and pausing. New actions can easily be added, due to the object-oriented structure of the system. Text output was implemented, but the low resolution of the HMD did not allow a useful amount of text to be included without hiding too much of the visible area; thus, digitized speech was used as the medium for explanation.

At any given time, one of these steps is current. When process advances into a new step, the associated actions are executed. If the user wasn't able to follow the actions on the first try (or rather wants to watch first before doing something) he can request a replay of the step actions via a voice command. If he gets lost, a context can be given by replaying the previous step and the current step; stepping backward is also possible.

**Rendering** The rendering is done using the three-dimensional models described in Section 4.3. The availability of data for the virtual door allowed us to incorporate the real door as an occlusion object, so that the virtual objects could be hidden by the real door (see Figure 14). This was done by rendering the virtual door into the $Z$-buffer of the graphics hardware, thus occluding objects rendered later without interfering with the contents of the frame buffer (which might be the background video image as used in feed-through mode). By using scan-line interleaved rendering and a straight-through video output, the system can send a separate image to each eye, resulting in a stereoscopic image which accurately conveys the depth involved.

**Code Reuse** The bulk of this application software was taken directly from the virtual reality system, which uses it for the same task (playing back prerecorded sequences of actions). It had to be ported to use Open-Inventor as a rendering engine, but the amount of code reuse allowed a fast and fluid realization of the system and an easy transfer of animation data.

## 5  Presentation

The system was demonstrated to the general public at the Hannover Industrial Fair in April 1998 as a part of the Fraunhofer Vision booth. To our knowledge, this was the first presentation of an AR application to a wide industrial audience in Germany. For the presentation, we used two independent but identical machines running the system; one was slaved to the other, so that both had the same user input and showed the same

**Figure 9.** Set-up at the Hannover Industrial Fair '98.

augmentations. One system was statically connected to the HMD camera and displayed its output on the HMD; the other was switchable between the HMD camera and a tripod-mounted external camera. It displayed its output on a large projection screen (see Figure 9).

The system ran for the whole week without problems. After an initial trial phase, the speech recognition was switched off in favor of automatically advancing animations. The ambient noise from talking robots and video screens in nearby booths proved to be too much for our microphone and speech recognition system.

Interested visitors were allowed to wear the HMD and take a peek at the application. Since, due to time constraints, no individual camera-to-eye calibration (see Section 4.2) was done, however, the results were not as overwhelming as some people expected.

Another unappreciated problem was the 200-ms initialization delay of the optical tracker. Right after donning the HMD, a number of visitors started to move their head around wildly, searching for the augmentations. The tracker never received a stable picture long enough to initialize. By asking the user to look at the squares and quickly switching to feed-through, we facilitated the initialization, whereupon the switch back to see-through conveyed the full, three-dimensional impression.

The general feedback of the audience was very positive. The physical presence of the door, in conjunction with the projection of virtual reality, provided an appropriate introduction to the concept of integrating computer-generated imagery into views of the real world. This was apparent after the demonstration was shown on German TV. In subsequent days, people came looking for our booth, some with very specific ideas and questions. A number of companies were interested enough to discuss concrete concepts of applications and projects, some of which are in the definition phase right now.

# 6   Results and Future Work

An augmented reality system was created for training the users to perform the doorlock assembly task. It uses CAD data taken directly from the construction and production database (as far as possible using today's hardware) and animation and instruction data that was prepared as part of a Virtual Prototyping planning session. This allows the system to be integrated into the existing infrastructure and to feed off data generated for other purposes. This gives added value to that data. We designed and implemented an optical tracking system using low-cost passive markers that is fast enough for careful HMD use.

Demonstrating the system at the Hannover Industrial Fair allowed a number of unrelated visitors to try it, and showed us the tracker is not quite able to handle naive users yet. We are looking into integrating inertial tracking hardware into the system, which should solve the problematic cases of fast head motion generating motion-blurred, and thus unusable, images for the optical tracker. At the same time, the optical tracker will be able to alleviate the weaknesses of the purely inertial system, namely its drift. The two approaches should complement each other nicely. The fast acceleration data provided by the inertial system will also be used for better motion prediction to compensate better for the system-immanent lag, improving registration between real and virtual objects.

Ambient sound turned out to be a bigger problem than we had expected; we are looking for more robust speech recognition software/hardware. Since the interface is extremely simple (ASCII strings of the recognized commands sent via RS-232), nearly any commercial solution should be easily attachable to the system.

One goal is to move the system onto a wearable platform like the Xybernaut machines [Xybernaut 98]. A Linux port is being developed for that purpose.

Industrial partners have been interested enough to start concrete project planning negotiations, so there seems to be a bright light on the horizon for industrial applications of augmented reality.

# 7   Acknowledgments

Thanks go to our colleagues Mike Weintke and Claudia Pilo for their hard work on the models right after a tough deadline. We also thank the BMW Virtual Reality Lab, especially Mr. Gomes de Sa and Mr. Baake for their support, for allowing us to use the CAD data, and for supplying a real door and doorlock for our demonstration. Parts of this work were supported by the European ACTS project CICC (AC017). Laboratory space and equipment were provided by ECRC.

**Figure 10.** Lever inside the door to be pushed to right position.

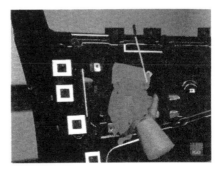

**Figure 11.** Gripping instruction.

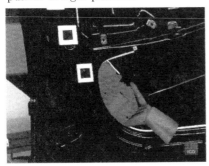

**Figure 12.** Moving instruction.

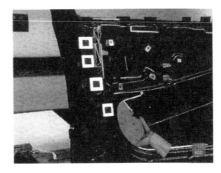

**Figure 13.** Problem area highlighting (antenna guidance).

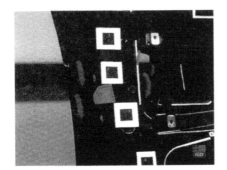

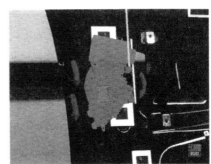

**Figure 14.** Screw-fixing instruction with and without occluding door.

# References

[Caudell 92]   T. Caudell and D. Mizell. "Augmented Reality: An Application of Heads-up Display Technology to Manual Manufacturing Processes," *HICCS '92,* 1992.

[Dai 96]   F. Dai, W. Felger, Th. Frühauf, M. Göbel, D. Reiners, and G. Zachmann. "Virtual Prototyping Examples for Automotive Industries," *Proceedings of Virtual Reality World (VRW '96),* 1996.

[Feiner 93]   S. Feiner, B. Macintyre, and D. Seligmann. "Knowlege-Based Augmented Reality," *Communications of the ACM,* Vol. 36, No. 7, pp. 53–61, 1993.

[Janin 93]   A. Janin, T. Caudell, and D. Mizell. "Calibration of Head-Mounted Displays for Augmented Reality Applications," *VRAIS '93,* 1993.

[Koller 97]   D. Koller, G. Klinker, E. Rose, D. Breen, R. Whitaker, and M. Tuceryan. "Real-time Vision-Based Camera Tracking for Augmented Reality Applications," *Proceedings of the ACM Symposium on Virtual Reality Software and Technology (VRST '97),* Lausanne, Switzerland, ACM Press, September 1997.

[Kutulakos 96]   K.N. Kutulakos and J. Vallino. "Affine Object Representations for Calibration-Free Augmented Reality," *Proceedings VRAIS '96,* Hong Kong, IEEE, pp. 1–20, 1996.

[Mann 98]   S. Mann. Repair instruction. http://lcs.www.media.mit.edu/projects/wearables/augmented-reality.html, 1998.

[Neumann 96]   U. Neumann and Y. Cho. "A Self-tracking Augmented Reality System," *Proceedings of Virtual Reality Software and Technology (VRAIS '98),* Hong Kong, pp. 109–115, 1996.

[Neumann 97]   E.A. Neumann. Personal Communication, August 1997.

[Starner 97]   T. Starner, S. Mann, B. Rhodes, J. Levine, J. Healey, D. Kirsch, R.W. Picard, and A. Pentland. "Augmented Reality Through Wearable Computing," *Presence, Special Issue on Augmented Reality,* Vol. 6, No. 4, pp. 386–398, August, 1997.

[State 96]   A. State, G. Hirota, D.T. Chen, W.F. Garrett, and M.A. Livingston. "Superior Augmented Reality Registration by Integrating Landmark Tracking and Magnetic Tracking," *Proceedings SIGGRAPH,* New Orleans, ACM, pp. 429–438, August 1996.

[Stricker 98]   D. Stricker. "A Fast and Robust Optical Tracker for Augmented Reality Applications," *Augmented Reality: Placing Artificial Objects in Real Scenes, Proceedings of IWAR '98,* A K Peters, Ltd., Natick, MA, pp. 129–145, 1999.

[Tsai 86]        R.Y. Tsai. "An Efficient and Accurate Camera Calibra-
                 tion Technique for 3D Machine Vision," *Proceedings CVPR*,
                 IEEE, pp. 364–374, 1986. See also http://www.cs.cmu.
                 edu/~rgw/TsaiCode.html.

[Webster 96]     A. Webster, S Feiner, B MacIntyre, W. Massie, and T. Krueger.
                 "Augmented Reality in Architectural Construction, Inspec-
                 tion, and Renovation," *Proceedings ASCE Third Congress on
                 Computing in Civil Engineering*, Anaheim, CA, pp. 913–919,
                 June 1996.

[Weng 92]        J. Weng, P. Cohen, and M. Herniou. "Camera Calibration with
                 Distortion Models and Accuracy Evaluation," *IEEE Trans. on
                 Pattern Analysis and Machine Intelligence*, Vol. PAMI-14, No.
                 10, pp. 965–980, 1992.

[Xybernaut 98]   Xybernaut Corporation, Xybernaut website, http://www.
                 xybernaut.com/

# Several Devils in the Details: Making an AR Application Work in the Airplane Factory

Dan Curtis, David Mizell, Peter Gruenbaum, and Adam Janin

**Abstract.** *Since 1990, Boeing scientists have been developing an application of augmented reality to aircraft wire bundle assembly. This paper describes the variety of detailed issues that they had to address in order to get a working prototype into the factory environment, and presents the results and conclusions from the factory pilot project run in July and August 1997.*

## 1   Introduction

This paper has a particular point of view. It discusses not only the augmented reality technology behind the wire bundle assembly application we have been developing at Boeing, but also all the unexpected technical problems and many of the non-technical problems we have had to solve in the process of implementing and testing this application. Moving technology as radically new as AR and wearable computers into a busy aircraft factory is not a simple undertaking. Every time you think you understand the requirements, you get surprised. Every time you think you have something ready to deploy, you encounter a new obstacle. Sometimes the obstacles are wearing suits.

## 2   Project History

Boeing's AR project has been going on for several years [Caudell 92] [Janin 93] [Sowizral 93]. It got started in early 1990, when Tom Caudell and David Mizell visited the Everett, Washington plant to get ideas from the aircraft engineers for VR projects. We ran into a manager who had just finished a job assignment as manager of the Everett factory wire shop. His question to us was: "What can you do about 'formboard?'"

He went on to describe the way in which an aircraft's electrical wiring is assembled into bundles for the sake of ease of installation. The assembly

47

**Figure 1.** Traditional aircraft wire bundle assembly "formboard"

takes place on one or more $3' \times 8'$ easels, called formboards. To assemble the bundle, the worker follows a schematic diagram glued to the front surface of the board. Pegs screwed into the board provide support for the wires while the bundle is being assembled. This approach implies that each formboard is more or less unique to the bundle that is assembled on it. This, in turn, implies considerable storage, construction, and transportation costs. A 747 has about a thousand wire bundles installed on it. Wire bundles are constantly being changed for different aircraft, since airlines want different combinations of engines, avionics, and passenger cabin layouts.

Tom immediately perceived that an AR system with a see-through display, a focal distance set at approximately arm's-length working distance, and a VR-style 6DOF head tracker could make the appropriate diagrams appear on a blank formboard, greatly reducing the wire shops construction and storage costs.

Later in the project, we discovered that these cost savings were actually secondary to the potential savings AR could provide for bundle forming time. Because there are hundreds of people forming wire bundles at Boeing, any significant productivity improvement has a big multiplication factor. Because we could put all the information the worker needed in the display–a feat not possible with traditional foamboard–we thought that we could realize a productivity increase. In the traditional system, some wire positioning information must be found in the "shop aid paper," 8.5" × 11" printouts that accompany the sets of wires that are to be assembled on the board. While workers dig through these. their attention must shift away from the workpiece. The AR system is capable of superimposing all this information on the formboard, obviating the need to refer to the shop aid paper.

# 3   General Technical Challenges

## 3.1   Head-Mounted Display

AR comes in two flavors these days. There is "optical see-through," where the user sees the real world through transparent or semi-transparent optics which also support the display of the computer-generated information. And there is "video see-through," where the user wears video cameras on his head, mounted as close to his eyes as is feasible. The user sees a video image of the real world in front of him, as captured by these cameras. The AR system is then able to incorporate, by video-mixing, computer-generated data with the camera images.

Video see-through was out of the question for the formboard application. Workers have to read the identification numbers laser-printed in a 6-point font on the insulators of the wires. We needed a head-mounted display with see- through optics. A prototype we had built for lab use around 1993 was binocular, but experience with it convinced us that a monocular apparatus was sufficient for this application, as long as the user could have the display over her dominant eye. Experience also convinced us that a monochrome VGA-resolution ($640 \times 480$) display was sufficient for this application.

## 3.2   Wearable PC

Wire bundle formers have to work in an area that can be rather large in one dimension, along the length of the formboards. Long wire bundles — for example, those that run the length of a 747 — are assembled on multiple-board runs, which can be 8 boards (i.e., 64 feet) long. This fact significantly influenced our system design. For one thing, tether cables of such a length would be unwieldy. We decided to use a system that was based on a computer worn by the user, in as self-contained a way as possible.

## 3.3   6DOF Head Tracker

The most difficult challenge that this length requirement placed on the AR system was developing an adequate head tracker. Commercially-available

6DOF trackers didn't meet the requirements. None had the range and accuracy that we needed. Most had to be tethered. Most were very vulnerable to the interference characteristics of a factory environment: lots of metal, lots of acoustic and RF noise. Cost was also a constraint. Factory managers wouldn't pay for an ultra-expensive tracking system, given that they would end up buying them for each of several hundred workers. We also perceived some psychological constraints. We became convinced that, even if we brought in a hundred physicists to sing in harmony that some laser-based head tracking system we implemented was eye-safe, the wire shop workers *still* didn't want a laser pointing at their eyes.

## 3.4   User Interface

We knew that we had to start with a clean slate to design the user interface. AR and wearable, hands-free computers represented a new HCI paradigm. The WIMP interface did not apply. It was only later in the project, however, that it was driven home to us how critical to worker productivity the user interface design was.

## 4   The 1997 Factory Pilot Project

From mid-1994 through 1997, our AR R&D work was accelerated greatly by matching support from DARPA's Technology Reinvestment Project. Our partners in the project, Carnegie Mellon University, Virtual Vision, Inc., and (for two years) Honeywell Military Avionics Division, built a series of AR system and wearable computer hardware prototypes for the team to test in realistic application environments. In the last year of this effort, we felt that we were far enough advanced with the AR technology to set up a small pilot project in one of Boeing's wire shops, with wire shop workers assembling bundles using augmented reality.

We tried to set up a fairly carefully designed experiment. About ten wire shop or mock-up shop workers participated. Some were very experienced at forming bundles. Others had worked in the wire shop and knew all the nomenclature, but had never done wire bundle assembly on the formboards. We chose, with wire shop expert help, one simple wire bundle, one of medium complexity, and one complex (two-board–length) bundle. We conducted 36 trials, with workers switching from one bundle to another and between the AR system and the traditional formboards in various sequences. We timed the workers on each task. We were concerned about eye fatigue, given that workers would be spending several hours with the AR optics over just one eye, so we had a Boeing optician perform visual

**Figure 2.** The TriSen, Inc. augmented reality system being used in the 1997 pilot project in Boeing's Everett, WA factory.

acuity tests at the beginning and again at the end of a working day several times during the pilot.

We took a couple of weeks to "shake down" the AR hardware and software and our experiment procedures. The experiment series itself took about 6 weeks.

Prior to the experiments, however, we had to develop a working AR system that was robust and capable enough to be handed over to wire shop workers and to help them do their jobs. In Section 5, we describe the work we and our partners had to do to make the system usable for the pilot project.

## 5 Making the Pilot Work

### 5.1 Hardware

By the fall of 1996, we were teamed with two vendors, each of which was developing a prototype AR system for us to use in the summer 1997 pilot project in the Everett wire shop. TriSen Systems, Inc., of Minneapolis, MN, developed an AR system based on videometric tracking. Their system uses a black $4' \times 8'$ aluminum board (fiducial board) on which has been painted an array of circular "fiducial markers," whose locations and radii are precisely known. A video camera mounted on the HMD photographs a portion of the board, and the image is analyzed to determine the location and size of the fiducials currently in view. The software then analyzes this pattern further and, from the visible fiducial pattern, determines the corresponding values of the six degrees of freedom of the HMD. Notable features of this system include:

- Highly stable and accurate tracking — Graphics vary in position by no more than 0.1".

- Rapid recovery of registration — If registration is lost–because, for instance, the user looks away from the board–then when the fiducials are again in the camera's view, the system reacquires the correct position automatically, usually in less than a second. In particular, this means that when the system is first started, there is no need to "initialize" the tracker by placing the HMD in some known position.

- Usability with multiple boards — The fiducial pattern can be extended in a non-repeating manner so that a board sequence of arbitrary length can be used. This is particularly important for the wire shop application.

- Fast, reliable calibration — TriSen has developed a reliable and easy-to-use calibration procedure.

Virtual Vision, Inc. of Redmond, WA developed an AR system using a tracking system provided by Intersense, Inc. of Cambridge, MA. This tracking system was a hybrid system, using a combination of an acoustic tracker and an inertial tracker to determine the six degrees of freedom of the HMD. The tracking system works as follows: An array of LEDs on the HMD fire, sending out an infrared pulse. A set of acoustic beacons, rigidly attached to an aluminum board on which the wire bundle is to be built, detect the infrared pulse, and each emits an ultrasonic pulse. An array of three microphones mounted on the HMD receives these ultrasonic pulses and registers the time of arrival. The signals are of different frequencies, so the system knows which beacon sent which pulse and is therefore able to use the time-of-flight information to compute the position and orientation of the HMD. The inertial system tracks changes of position between arrivals of the various acoustic pulses, and its information is combined with the acoustic information to compute final 6DOF values. Though this system was not ready in time for the pilot, the concept has a number of interesting characteristics. For example:

- It is not necessary to have a fiducial board with a fiducial pattern painted on it. The acoustic beacons must be placed in precisely known locations, but there are only a few of them, and they do not have to be located in a planar array. The beacons could, in principle, be placed in known locations in the interior of an airplane fuselage, so that the AR system could be used for actually installing wire bundles or other equipment in the airplane. In order to make this feasible, it is necessary to have a quick and effective procedure for calibrating the system.

• Beacons can be placed so that they are not occluded by the workpiece. The videometric system can have problems tracking if a significant fraction of the fiducial pattern is covered up. We tried to include in the pilot project a very complex, 800-wire bundle, but the videometric tracker lost track once the worker got approximately half of the wires on the board. There was a section in the center of the board which was very dense with wires and the paper tags with which some of the wires are marked. When the worker leaned down to work in that section, the tracking system couldn't find any fiducial marks through all the clutter.

Both the TriSen and the Virtual Vision/InterSense systems use a specially-built aluminum board in place of the conventional formboard. Quarter-inch holes are drilled in the boards so that they resemble sheets of pegboard. The holes are used for mounting the pegs used to support the wire bundles while they are on the board. For the pegs, we have been using "Cleeco" fasteners, temporary fasteners which are widely used in aircraft construction. In TriSen's case, the dimensions of each board are $4' \times 8'$. The designers wanted to leave a foot of room at the top of the board which they knew would be unobstructed by wires. The TriSen board is painted with the array of fiducial polka dots used by the videometric tracker. The Virtual Vision board is $3' \times 8'$ and holds two rows of acoustic transmitters mounted on 3" stems, to keep them above the wires and other equipment being placed on the board.

## 5.2   Software

### Data Translation

One of the largest technical obstacles was the translation of data from the existing formboards to the AR data format. Data for forming wire bundles is held in two separate places: the mostly geometric data for the formboard diagrams in the aircraft CAD system, and the wire run information contained in an internally developed, ancient COBOL database. The form-board data describes the 2D paths on which the wires are bundled as well as any information that that pertains to points in 2D space (for example, "Place wire ties every 2 inches from here to there"). The wire run data describes which wires run between which connectors and any special information that pertains to individual wires (for example, "wire 43-20 should be fastened to lug using specification BAC-xyz").

To generate the formboards, Boeing runs its 3D wire paths through software that flattens them to 2D while maintaining lengths. After this, various annotations are added manually. Once in this format, the form-board becomes nothing more than a series of unordered line segments and text in 2D space, and it is very difficult for software to parse. A small amount of information is placed in a "hidden" layer of the data structure that allows the paths between connectors to be determined. By comparing this information with the data from the wire run database, it is possible to find the wire path for each wire in the bundle. However, we wanted the AR system to show only the annotations that were relevant for the wires that were being worked on at the time. Unfortunately, there was no information other than proximity to determine which annotations belonged to which wire paths. Therefore, any segment or text that was within a given distance from a wire path (about one to two inches) was considered to be associated with the wire path. In the end, manual clean-up had to be done on the data for each of the bundles in order to make sure the correct line segments and text were in each "world." This clean-up ranged from a few hours to several days, depending on the complexity of the bundle.

## Preferred Process

A "preferred process" for carrying out a task consists of a sequence of sub-tasks which, when carried out in a prescribed order, accomplish the given task in a way that is, in some appropriate sense, optimal. The de-composition into sub-tasks should be sufficiently fine-grained that possible variation in procedure within each sub-task is reduced to an acceptable level. Having a preferred process for a task reduces variability in the com-pleted work and can be a significant aid in training workers. It has long been a goal of the Boeing wire shops to implement such a preferred process for the task of forming wire bundles. Currently, Boeing wire shop work-ers are presented with a full-sized, diagram of the wire bundle (printed on paper, and then glued onto a $3' \times 8'$ plywood formboard), together with the "shop aid" paperwork detailing the content of the bundle (wire types, wire gauges, connectors, etc.). There is no reasonable way of enforcing a requirement that the bundle be assembled in a particular order. For the pilot project, a preferred process was agreed upon after discussions with wire shop personnel. This process calls for the wires in a given "wire group" (usually consisting of all the wires emanating from a particular connector) to be routed together until a "breakout point" (where branching occurs) is encountered. Then the system tells the worker which branch to follow first, the decision being based on some agreed-upon criteria. This represented the first time the wire shop had ever had a specification of the preferred bundle assembly process that was precise and general enough to be implemented into our translation software. We had to come up with a set of complete and consistent rules for deriving an appropriate sequence from the bundle geometry and connectivity information we were given.

## User Interface

The AR user interface uses a set of files, called "worlds," each of which contains part of the total information about the bundle. For instance, suppose a world is loaded which contains a certain set of wire paths and associated sleeves and flag note indicators. A second world might contain the text of a flag note (special information about how a certain wire or group of wires is to be handled) appearing in the first world. The AR system has a single input device: a one-button clicker which is used to select text items appearing in the HMD display. The user sees a box drawn around that piece of text which is nearest to the center of the display, and if the button is single-clicked, a new world file, to which the text item was linked, is loaded and displayed. For example, suppose the world mentioned above contains the flag note symbol "F10." If the user moves her head until the F10 has a box around it and then single-clicks the button, the world containing the text of the flag note is loaded and the user sees this text displayed. Once the note has been read, the user can double-click and the previous world file is reloaded. Use of links in this manner allows the AR system to eliminate the need to consult paperwork. Information, such as the contents of flag notes, sleeve notes, and wire notes, can now be accessed through links to world files containing the relevant information. The AR systems user interface is well-suited to implementation of a preferred build process as described above. The key to this suitability is the fact that display of the AR system does not have to show the worker the entire bundle at once. Once the build process has been defined as a sequence of well-defined tasks, the AR system software can be configured so that, at any given stage of the build process, the display shows only what is necessary to complete the "current task." This sequence of tasks and the corresponding views presented to the user constitute a preferred build process, and it is enforceable since the user is shown only the information needed to complete the current task.

### The Compass

Because the HMD only shows a small window into an AR world, early versions of the interface resulted in a problem when moving from one world to the next. A wire path might lead a user to one corner of the board, but when the user clicked to move onto the next wire (and into the next "world"), everything that needed to be displayed might be several feet away, and the user would see nothing but a blank screen. The user would then waste time scanning the board to find where to begin work again. In order to guide the user, an AR "compass" was created for the bottom corner of the screen. Each world has a starting point in 3D space. This point is projected onto the virtual screen, and then the compass (consisting of an arrow with a circle around it) points to that spot on the virtual screen. The user need only follow the arrow and eventually the starting point will come into view.

## 5.3   Miscellaneous Hiccups and Obstacles

A production factory floor is not always the ideal test environment for a new technology. The pilot study was conducted during one of the busiest times in the wire shop's history. The pilot study was given low priority compared to manufacturing demands, and this both limited and delayed the number of wire kits that could be used to assemble each bundle. As a result, we tore down and reused the wire kits after each test. Workers were on mandatory 12-hour shifts, and although the wire shop kept its commitment to the study by making sure that the workers were available, they were often exhausted. Another factor that may have skewed our timing measurements was the "curiosity factor." Workers wearing the AR unit were likely to be visited by curious coworkers asking what was going on, and often wanting to try out the HMD. We attempted to keep such visits short, but they occurred with some frequency.

On the other hand, some potential problems did not materialize. Some of the subjects spoke English as a second language (which is common among workers in the wire shop), but this did not prove to be an obstacle either to training or to obtain feedback. Although the area of the wire shop where the tests were conducted was relatively quiet when the site was chosen, there was a period of time when a mock-up airplane was being dismantled nearby, and the noise from the construction was sharp and loud. We feared for the performance of the acoustic tracker, but it showed no signs of being affected.

# 6   Results of the Pilot

The first highly positive result that came out of the pilot study was that the concept worked. Using AR, workers were able to build simple and moderately complex bundles with only a brief training session. After they had been formed, two of the bundles were removed from the AR pegboards and re-positioned on their standard formboards. The fit to the original paths was close enough that a Boeing Quality Assurance worker asserted that they would have passed inspection had they been formed in a traditional fashion.

The data collected during the pilot project concerning bundle forming time showed a statistical dead heat between the current method and the proposed AR method. This surprised us, since we expected that elimination of searching through paperwork would result in substantial speed-ups in the bundle forming task. Much of the reason for the lack of speed-up can be traced to the user-interface, which forced the worker to search repeatedly through a set of wires in order to find the subset of them which branch in a particular direction at a breakout point. Preferred process strategies other than the one used for the pilot may eliminate this problem. It may be possible, as a preprocessing step, to efficiently decompose each wire group into a set of "sub-groups" so that the wires in a given sub-group have a common termination point. If this were done, then a viable strategy would consist of routing these sub-groups, one at a time, from beginning to end. This would completely eliminate searching for particular wire numbers. Preliminary studies indicate that such strategies can result in large productivity gains.

The results of the eye tests before and after using the AR system were reassuring, in their own way. *Everyone* got slight amounts of eyestrain — people using the AR system and people using the traditional formboard. Apparently, spending an eight-hour shift reading wire IDs printed in 6-point fonts will give anyone eyestrain, whether the worker is using AR or not. There was no significant difference between the eyestrain measured on the AR users and that measured on the traditional formboard users.

Reactions to wearing the computer and the HMD were a rather bipolar mixture. Several people wore the system for a full shift several different days, and never complained of any discomfort. Other people complained of headaches, neck aches, etc. within minutes of putting the system on.

The contrast was so great that we suspect that the real discomfort experienced by this second group was psychological. The TriSen HMD is not a fashion statement, and it is not hairdo-friendly. The work area we were allocated for the pilot project was right next to the mock-up shop break room, and participants had to expect all their friends to come by at lunch or break time, and see them looking like they'd been assimilated by the Borg. There are significant attitudinal differences between MIT Media Lab grad students who volunteer to wear computing equipment all day and machinists' union members who are *asked* by their supervisor to wear it.

## 7   Next Steps

It was clear to all of us who participated in the 1997 pilot project that the technology was not yet ready for large-scale deployment in the Boeing wire shop. The user interface in some ways slowed the user down instead of helping her speed up. The ergonomics and overall comfort of the wearable equipment needed to be improved before it would be acceptable to many of the workers.

We are now engaged in what we think is the most important step towards improving the user interface. We are integrating speech recognition software into it. Reliable speech recognition technology, we believe, will enable us to accelerate greatly the wire bundle forming process. In particular, we can use it to implement what we have named the "sort vs. search" paradigm. In the 1997 user interface implementation, when the user was routing wires at a breakout, he had to click the mouse button to bring up a list of wire numbers that took a particular branch, and then repeatedly search until he had found each of those wires in the group he was routing at the time. With speech recognition, the worker will be able to select any wire in the group, read its ID number into the AR system, and immediately be told which way to route it. He will, in other words, be able to route all the wires in the group past that branch point in a single pass, rather than by repeatedly searching through the group.

It turns out that this change in the software requires a change in the hardware. TriSen wrote its videometric tracking system to run under DOS and doesn't plan to change it. For us to run state-of-the-art speech recognizers, we need to run the application and the user interface code under a newer operating system. TriSen has agreed to add a processor to the system that is dedicated to running their tracker code. It will continue to run DOS. We will run the user interface and the app on the "main" processor, under whatever OS we choose. We are developing our first prototype under Windows NT.

Other hardware developments that will improve the next-generation AR system are happening simply because the industry is moving in that direction. Wearable computers are rapidly becoming faster, smaller, and lower-power. Increased processing speed is significant in two ways. First, it enables real-time responsiveness of the current generation of commercially-available speech recognition software. Second, it reduces the "swimming" of the graphics image when the user moves her head, since faster processing reduces overall system latency. The HMDs are also becoming smaller, lighter, and more comfortable. See-through HMD prototypes already exist which use an eyeglasses format [Spitzer 97]. Boeing's factory workers are required to wear safety glasses anyway, so once we can obtain see- through HMDs that resemble (and qualify as) safety glasses, the acceptability problem will largely be solved.

Finally, we are eager to develop or promote the development of alternative 6DOF head tracking technologies. The TriSen videometric tracking technology works quite well for the wire shop application, where it is reasonable to use the formboard with its fiducial pattern. There are, however, many other potential applications of the concept of augmented reality where use of the formboard is not feasible. For example:

- Installation of wire bundles (and other equipment) in the airplane

- Training of workers

- As an aid to carrying out complex maintenance tasks.

Since we cannot readily bring a fiducial board into small interior spaces in an airplane, and since Boeing has rejected the idea of painting all the surfaces of the airplane with polka dots, it seems that an alternative to videometric tracking is needed. For objects designed on a modern CAD system, there are often specific, readily identifiable features on the object, whose locations are precisely known in CAD-coordinates. Sensors, such as the acoustic beacons mentioned above, could be mounted at such locations and used to track head motion.

## 8    Conclusions

For some people in the research community, a new computing concept is "done" as soon as you've written a paper about it. A few stop sooner than that, but they usually don't get tenure. For others, "done" consists of implementing the first demo. They apparently tend to gravitate into jobs in the research funding agencies. For us, "done" means that there are substantial numbers of Boeing wire shop workers using AR systems every day to assemble production wire bundles. We aren't there yet, but we believe that we're within a year or two of getting there.

We have tried to give the reader a flavor of the wide variety of obstacles and problems you are faced with when this third choice is your definition of "done." Many of the obstacles are human. One of our colleagues jokes about the "Boeing immune system" that swings into action when you try to introduce a new technology: bureaucrats who consider all computers wasteful, people who have a competing technical approach, people who resist any new technology for fear that it might expose their ignorance of it, and people who'd sincerely like to help you except that their time is consumed by getting airplanes out the door.

Other problems are technical, but unexpected. We didn't know about the wire shop managers' desire to implement a "preferred process" for bundle assembly. That was both serendipitous and a software design challenge. Others are not technically profound, but very messy — such as extracting the data we needed to display from legacy databases. It took us weeks to find someone who knew the format of the old data.

There is a rule of thumb in the R&D community which says that it takes 10 years to transition a new idea from original concept to production deployment. We seem to be right on schedule. We hope that this paper helps illustrate why it takes that long.

# References

[Caudell 92]   T. Caudell and D. Mizell. "Augmented Reality: An Application of Heads-Up Display Technology to Manual Manufacturing Processes," *Proceedings, Hawaii International Conference on Systems Sciences,* Kauii, Hawaii, IEEE Press, pp. 659–669, January 1992.

[Janin 93]   A. Janin, D. Mizell, and T. Caudell. "Calibration of Head-Mounted Displays for Augmented Reality Applications," *Proceedings, Virtual Reality Annual International Symposium,* Seattle, WA, IEEE Press, pp. 246–255, September 1993.

[Sowizral 93]   H. Sowizral and J. Barnes. "Tracking Position and Orientation in a Large Volume," *Proceedings. Virtual Reality Annual International Symposium,* Seattle, WA, IEEE Press, pp. 132–139, September 1993.

[Spitzer 97]   M. Spitzer, N. Rensing, R. McClelland, and P. Aquilino. "Eyeglass-Based Systems for Wearable Computing," *Proceedings, First International Symposium on Wearable Computers,* Cambridge, MA, IEEE Computer Society, pp. 48–51, October, 1997.

# Part II

# Novel User Interface Paradigms

*The widespread use of augmented reality technology will radically change the way we access and present information. The short papers and position statements in this section give a glimpse of the new human-computer interface modalities that will become increasingly common.*

# Spatially Augmented Reality

Ramesh Raskar, Greg Welch, and Henry Fuchs

**Abstract.** *To create an effective illusion of virtual objects coexisting with the real world, see-through HMD-based augmented reality techniques supplement the user's view with images of virtual objects. We introduce here a new paradigm, Spatially Augmented Reality (SAR), where virtual objects are rendered directly within or on the user's physical space.*

*A key benefit of SAR is that the user does not need to wear a head-mounted display. Instead, with the use of spatial displays, wide-field-of-view and possibly high-resolution images of virtual objects can be integrated directly into the environment. For example, the virtual objects can be realized by using digital light projectors to "paint" 2D/3D imagery onto real surfaces, or by using built-in flat panel displays.*

*In this paper we present the rendering method used in our implementation and discuss the fundamentally different visible artifacts that arise as a result of errors in tracker measurements. Finally, we speculate about how SAR techniques might be combined with see-through AR to provide an even more compelling AR experience.*

## 1 Introduction

In *Spatially Augmented Reality (SAR)*, the user's physical environment is augmented with images that are integrated directly into the user's environment, not simply into their visual field. For example, the images could be projected onto real objects using digital light projectors, or embedded directly in the environment with flat panel displays. For the purpose of this paper, we will concentrate on the former. While the approach has certain restrictions, it offers an interesting new method for realizing compelling illusions of virtual objects coexisting with the real world. The images could appear in 2D, aligned on a flat display surface, or they could be 3D and floating above a planar surface, or even 3D and floating above an irregular surface.

For 2D non-head-tracked SAR, the images representing virtual objects do not continuously change with user motion. For example, in the Luminous Room [UnderKoffler 97], the user's environment is enhanced with synthetic images projected on flat surfaces. However, the user can be head-

63

tracked and the images updated dynamically in order to create the illusion that virtual objects are registered to real objects. Shuttered glasses can be used to facilitate stereo imagery, further enhancing the 3D effect of the virtual imagery. For the purpose of this paper, we will focus on technologies for head-tracked SAR where virtual objects are rendered on irregularly shaped real objects. While not appropriate for every application, this method does not require the user to wear a head- mounted display. Instead, with the use of spatial displays, wide-field-of-view and possibly high-resolution images of virtual objects can be integrated directly into the environment.

It is our work on the Office of the Future [Raskar 98a] that led to the realization and implementation of the SAR paradigm as described and analyzed in this paper. Specifically, when exploring the use of irregular (non-planar) surfaces for spatially immersive displays, we realized that the registration problem becomes the somewhat unusual one of having to register 2D imagery with 3D physical geometry. This is similar to conventional AR applications, except for one crucial point: the 2D imagery exists on a virtual image plane that is attached to the (fixed) physical display's surface instead of being attached to the user's (moving) head. However, as with conventional AR techniques, we realized that what is most important in terms of results is what the *viewer* sees. In retrospect it makes sense, but it was only after some further thinking that we recognized the fundamentally different manifestation of visual registration error in this paradigm. In a similar manner to the situation in a CAVE (spatially immersive display) [Cruz-Neira 93], such error is virtually unaffected by viewpoint orientation error, and viewpoint position error results in a form of image shear rather than a simple mis-registration.

The occlusion relationships in SAR are also different from those in see-through AR systems. In SAR, a real object can occlude the virtual object. Thus, for example, bringing your hand in front of your face will occlude the virtual object behind it, thereby maintaining the illusion that the virtual object exists in the real world. On the other hand, a virtual object cannot obstruct the view of a real object even if it is intended to float in front of that object.

User interaction with virtual objects in SAR, which involves tracking multiple parts of the user's body, such as head and hands, in 3D, is dependent on the accuracy of tracking devices. In head-mounted-display virtual reality (HMD-VR), where a virtual interface device manipulates the virtual world, small tracking measurement errors are not noticeable. In SAR, however, accurate spatial relationships among the user's head, the interacting body part, and the virtual object need to be maintained. Thus, the user interaction issue in SAR is a subset of registration issues that arise in AR.

SAR techniques could be used for many different applications. For example, architectural designing applications could benefit from the perceived ability to visually modify portions of a real physical environment of tabletop architectural models. Similarly, engineers could use the approach to "paint" alternate appearances on or inside a life-sized automobile mockup. The approach could also be used for product training or repair: one could set the product in the SAR environment and have the system render instructions directly on the product. Doctors could use the SAR to visualize and discuss virtual information that is projected onto a patient or mannequin, while simultaneously visualizing remote collaborators whose imagery and voices are spatially integrated into the surrounding environment.

## 2   Previous Work

Various levels of integration of virtual objects with users' physical environments are seen in current augmented reality (AR) systems [Milgram 94a]. HMD-VR has been widely used to generate synthetic images inside head-tracked, head-mounted displays that occlude the view of the real world but give the illusion of spatial and temporal context in the user's physical world. Optical and Video See-through Augmented Reality (OST-AR and VST-AR) systems combine the real scene viewed by the user and a virtual scene generated by the computer in order to augment the view with additional information [Milgram 94b] [State 96].

Some systems have integrated synthetic images with real scenarios for a static user. Dorsey et al. provide a useful framework in the context of theater set design [Dorsey 91], where a pre-distorted image appears correct when projected onto a curved backdrop of the theater. Luminous room [UnderKoffler 97] is a partially immersive spatially integrated environment. The system projects and then generates 2D images on flat surfaces in a room in order to enhance the user's environment. The HoloGlobe exhibit uses High Definition Volumetric Display, along with precision optical components such as parabolic mirrors and beam splitters, to display huge amounts of data concerning global change on a four-foot, 3D floating image of the Earth [Hologlobe]. Viewers can walk around the 3D image and see it from different angles. The users do not need to wear any gear. The Office of the Future (OOTF) [Raskar 98a] places users "inside", rather than outside, the augmented reality, surrounding them with synthetic images such as those generated by spatially immersive displays (SID). CAVE [Cruz-Neira 93] and ARC's dome-shaped displays [Bennett 98] are other examples of SID. However, in OOTF, the display surfaces are not limited to the designated flat walls (or parameterized surfaces) — they may be everyday surfaces.

# 3   Methods

To create the illusion, for a moving user, that virtual objects are registered to real objects, we need to know the position of the user, the projection parameters of the display devices, and the shape of the surfaces of the real objects in the physical environment; we must also be able to render virtual objects on those surfaces. Here, we will describe a method for each of these steps as implemented by a unified projector-based SAR system.

## 3.1   Display Surface Shape Extraction

The 3D surface shape extraction can be achieved using a calibrated [Tsai 86] projector-camera pair; structured light patterns are projected and observed by the camera. [Raskar 98c] describes a near-real-time method for capturing the 3D shape of the display surface, and [Raskar 98a] describes a unified approach to capturing and displaying images on irregular (non-planar) surfaces.

## 3.2   Rendering and Viewing Method

The process of projecting images on irregular surfaces so that they appear correct to a static user has been described in [Dorsey 91], [Max 91], [Jarvis 97], and [Raskar 98d]. In [Raskar 98a], a real time technique was introduced for generating such images for a moving head-tracked user. Here, we will describe how it can be used for SAR even when not all the viewing parameters are known. Let $V = V_i * V_e$ represent the user's-eye-perspective projection matrix, where, $V_i$ is the projection matrix and $V_e$ is the transformation matrix (subscripts $i$ and $e$ are used for intrinsic and extrinsic parameters, respectively). $E = E_i * E_e$ is an intermediate projection matrix, which shares the center of projection (COP), with the user's eye, and $P$ represents the projector's perspective projection matrix. Display surface model is $D$, and $G$ is the graphics model we wish to render. We will use the notation $I(r, g, b, z) = M * [G]$ to indicate that image $I$ is generated from 3D colored model $G$ using perspective projection matrix $M$ (using any rendering method). Note that $M^{-1} * I(r, g, b)$ represents a set of 'colored' rays, and $M^{-1} * I(r, g, b, z)$ represents a colored surface interpolating 3D points due to termination of the rays at the corresponding depth. We want to present an image $I = V * [G]$ to the user with *only* the COP for $V$ known. (Note that in general it is difficult to estimate correctly the COP for $V$, due to the obvious difficulty in co-locating the tracking sensor at the human eye COP. However, as we point out below, erroneously estimating the COP results in a fundamentally different visual artifact than that generated by head-mounted AR, an artifact that would be less noticeable in typical applications.)

**Rendering**

Step I: (a) Compute the desired image color by rendering $G$:

$$I1(r, g, b) = E * [G]$$

(b) Update only the depth buffer (without overwriting color) by rendering the display surface model $D$.

$$I1(z) = E * [D]$$

Step II : Apply a 3D warp to image $I1$ to effect the view transformation from the projection system defined by $E$ to the projection system defined by $P$. This is similar to the process involved in image-based rendering of images with depth. (This can also be achieved by projective texture mapping: project texture $I1(r, g, b)$ onto $D$ and render using $P$ as described in [Raskar 98b]).

$$I2(r, g, b, z) = P * E^{-1} * I1(r, g, b, z)$$

**Viewing**

When we are viewing in the real world, if $I2$ is projected by the projector onto the display surface and viewed by the viewer, we get image $I3$:

$$
\begin{aligned}
I3(r, g, b) &= V * [D], \text{where } D \text{ is now display surface} \\
&\qquad\qquad \text{colored by projected light} \\
&= V * P^{-1} * I2(r, g, b, z) \\
&= V * E^{-1} * I1(r, g, b, z)
\end{aligned}
$$

Note that, although $E^{-1} * I1(r, g, b, z)$ is not the same as $[G]$, because the depth values in $I1$ are due to display surface model $[D]$, $E^{-1} * I1(r, g, b)$ represents 'colored' rays due to $[G]$'s meeting at COP of the projection system of $E$. Thus, if COP for the systems represented by $V$ and $E$ are the same, I3 is the intersection of the colored rays with the image plane for $V$. This is the perspective projection of $[G]$ such that $I3(r, g, b) = V * [G]$.

Repeating this for the second eye with a different $E$ (i.e., a different COP) allows one to create stereo images. In practice, $E$ is chosen to approximately match the parameters (especially field of view and orientation) of the user's eye, in order to minimize the size of the required frame buffer for rendering.

## 3.3   Registration Artifacts

For purely virtual environments, it suffices to know the approximate position and orientation of the user's head in a fixed world-coordinate system. Small errors are not easily discernible, because the user's visual sense tends to override the conflicting signals from his or her vestibular and proprioceptive systems. However, in see-through AR or SAR, preserving the illusion that virtual and real objects coexist requires proper alignment and registration of virtual objects to real objects [Azuma 94]. Traditional AR methods use a body-centric coordinate system to render synthetic objects, and SAR methods use a fixed world-coordinate system to render them. However, in both, the registration errors are caused by a number of factors such as system delay, optical distortion, and tracker measurement error, and are difficult to correct with existing technology [Holloway 95].

The tracking requirements for registration in SAR are similar to those in SID-VR systems, because real and virtual objects lie in the same fixed world-coordinate system. Thus, static calibration errors can play an important role in registration. They include incorrect estimate of transformations between display devices, tracker, and world-coordinate system. In video see-through AR, optical distortion in camera lenses introduces error when computer-generated imagery is combined with video images [State 96]. On the other hand, registration error in SAR is introduced by optical distortion in the projector lenses.

Let's compare the ways in which errors in the actual and estimated perspective projection parameters of the system result in visible artifacts. In see-through AR, such errors result in virtual object 'swimming' with respect to the real objects [Holloway 95]. In SAR, these errors lead to fundamentally different types of visible artifacts.

As noted in the previous section, the system requires only the location of user's eye and not the orientation in order to render perspectively correct images of virtual objects on real surfaces. Hence, change in only the orientation of the user does not change the way real objects are 'painted.' Similarly, an error in the measurement of the orientation of the tracker will not result in misregistration error. An error in the measurement of the position of the user will result in shearing of virtual objects that are projected onto the real surfaces with which they are expected to be registered. The shear visible at any point on the virtual object is proportional to the distance between the virtual point and the real display surface on which the object is projected. Therefore, a virtual object not in 'contact' with a real surface but fixed in the real world will exhibit shear as well as swimming when there is a position measurement error.

## 4   Advantages of SAR

A key benefit of SAR is that the user does not need to wear a head-mounted display. In [Bryson 97], the authors note various advantages of spatially immersive displays over head-mounted displays. SAR shares similar benefits. In SAR, large-field-of-view images can be generated with a greater amount of integration of virtual objects with the real world, and the sense of immersion can be improved. Projector-based SAR may well allow higher-resolution and brighter images of virtual objects, text, or fine details. Since virtual objects are typically rendered near their real-world locations, eye accommodation is easier.

## 5   Problems with SAR

The most crucial problem with projector-based SAR is its dependence on display surface properties. A light-colored diffuse object with smooth geometry is ideal. It is practically impossible to render vivid images on highly specular, low-reflectance, or dark surfaces. The ambient lighting can also affect the contrast of the images. This limits the application of SAR to controlled lighting environments, and restricts the types of objects with which virtual objects may be registered. For front-projector-based SAR, shadows of the user can create problems. This can be partially overcome using multiple projectors. SAR also allows only one active head-tracked user in the environment at any instant, because the images are created in the physical environment rather than in individual user space. Time multiplexed shuttered glasses can be used to add more users who are active and head-tracked.

## 6   Future Work

We have built a proof-of-concept system and demonstrated the SAR ideas in our laboratory; however, we are anxious to test the approach with a real application — for instance, one of the many applications listed in the introduction, or an application that involves a more complicated physical display surface, e.g., room-sized terrain visualization.

While the majority of our efforts to date have been focused solely on projector-based SAR, a hybrid environment could be built with, for example, digital light projectors and a see-through HMD. While this would require the user to wear an HMD, we believe that such a hybrid has the potential to offer the best of both worlds, combining all of the advantages of conventional see-through AR and this new SAR paradigm. As we gain experience building both conventional AR systems and projection-based SAR systems, we look forward to building and experimenting with such a hybrid environment.

## 7   Conclusion

Others have certainly used light projectors and even LCD panels to add virtual imagery to real environments. However, this is essentially an augmented reality problem, albeit an unusual one, and we are excited about the opportunities that accompany this realization. On the one hand, some of the conventional wisdom about AR can be applied to address certain SAR problems, and on the other hand, SAR can be used to address some difficulties with conventional AR for some applications. We look forward to refining our ideas and the related algorithms, analyzing further the relevant error sources, and pursuing some of the many applications we have in mind.

## References

[Azuma 94]      R. Azuma and G. Bishop. "Improving Static and Dynamic Registration in an Optical See-through HMD." *Proceedings of SIGGRAPH '94 Computer Graphics*, Orlando, FL, pp. 197–204, July 1994.

[Bennett 98]    D.T. Bennett, Chairman and Co-Founder of Alternate Realities Corporation, 215 Southport Drive, Suite 1300, Morrisville, NC 27560, USA. Cited March 29, 1998, available at http://www.virtual-reality.com.

[Bryson 97]     S. Bryson, D. Zeltzer, M.T. Bolas, B. de La Chapelle, and D. Bennett. "The Future of Virtual Reality: Head Mounted Displays Versus Spatially Immersive Displays," *SIGGRAPH '97 Conference Proceedings*, Addison-Wesley, pp. 485–486, August 1997.

[Cruz-Neira 93]   C. Cruz-Neira, D.J. Sandin, and T.A. DeFanti. "Surround-Screen Projection-Based Virtual Reality: The Design and Implementation of the CAVE," *SIGGRAPH '93 Conference Proceedings*, Addison Wesley, pp. 135–142, 1993.

[Dorsey 91]   J.O'B. Dorsey, F.X. Sillion, and D.P. Greenberg. "Design and Simulation of Opera Lighting and Projection Effects," *SIGGRAPH '91 Conference Proceedings*, Addison-Wesley, pp. 41–50, 1991.

[Holloway 95]   R. Holloway. "Registration Errors in Augmented Reality Systems," PhD Thesis. University of North Carolina at Chapel Hill, 1995.

[Hologlobe]   (Cited July 10, 1998) http://www.si.edu/hologlobe/welcome.htm

[Hornbeck 95]   L.J. Hornbeck. "Digital Light Processing for High-Brightness High-Resolution Applications," cited April 21, 1998. Available from http://www.ti.com/dlp/docs/business/resources/white/hornbeck.pdf, 1995.

[Jarvis 97]   K. Jarvis. "Real Time 60Hz Distortion Correction on a Silicon Graphics IG," *Real Time Graphics*, Vol. 5, No. 7, pp. 6–7, February 1997.

[Max 91]   N. Max. "Computer Animation of Photosynthesis," *Proceedings of the Second Eurographics Workshop on Animation and Simulation*, Vienna, pp. 25–39, 1991.

[Milgram 94a]   P. Milgram and F. Kishino. "A Taxonomy of Mixed Reality Visual Displays," *IEICE (Institute of Electronics, Information and Communication Engineers) Transactions on Information and Systems, Special issue on Networked Reality*, Vol. 12, No. E-D, pp. 1321–1329, December 1994.

[Milgram 94b]   P. Milgram, H. Takemura, A. Utsumi, and F. Kishino. "Augmented Reality: A Class of Displays on the Reality-virtuality Continuum," *SPIE Vol. 2351-34, Telemanipulator and Telepresence Technologies*, pp. 282–292, 1994.

[Raskar 98a]   R. Raskar, G. Welch, M. Cutts, A. Lake, L. Stesin, and H. Fuchs. "The Office of the Future: A Unified Approach to Image-Based Modeling and Spatially Immersive Displays," *SIGGRAPH '98 Conference Proceedings*, Addison-Wesley, pp. 179–188, July 1998.

[Raskar 98b]        R. Raskar, M. Cutts, G. Welch, and W. Stürzlinger. "Efficient Image Generation for Multiprojector and Multisurface Displays," UNC Computer Science Technical Report TR98-016, University of North Carolina at Chapel Hill, March 1998.

[Raskar 98b2]       R. Raskar, M. Cutts, G. Welch, W. Stüerzlinger. "Efficient Image Generation for Multiprojector and Multisurface Displays," *Rendering Techniques '98*, G. Drettakis and N. Max, *Proceedings of the Eurographics Workshop* Vienna, Austria, June 29–July 1, 1998.

[Raskar 98c]        R. Raskar, H. Fuchs, G. Welch, A. Lake, and M. Cutts. "3D Talking Heads: Image Based Modeling at Interactive Rate using Structured Light Projection," UNC Computer Science, Technical Report TR98-017, University of North Carolina at Chapel Hill, 1998.

[Raskar 98d]        R. Raskar, G. Welch, and H. Fuchs. "Seamless Projection Overlaps Using Image Warping and Intensity Blending," *Fourth International Conference on Virtual Systems and Multimedia*, Gifu, Japan, November 1998. http://www.cs.unc.edu/~raskar/Office/

[State 96]          A. State, G. Hirota, D.T. Chen, W.F. Garrett, and M.A. Livingston. "Superior Augmented Reality Registration by Integrating Landmark Tracking and Magnetic Tracking," *Proceedings of SIGGRAPH '96. Computer Graphics*, New Orleans, pp. 429–438, August 1996.

[Segal 92]          M. Segal, C. Korobkin, R. van Widenfelt, J. Foran, and P.E. Haeberli. "Fast Shadows and Lighting Effects using Texture Mapping," *SIGGRAPH '92 Conference Proceedings*, Addison Wesley, Vol. 26, pp. 249–252, July 1992.

[Tsai 86]           R.Y. Tsai. "An Efficient and Accurate Camera Calibration Technique for 3D Machine Vision," *Proceedings of IEEE Conference on Computer Vision and Pattern Recognition*, Miami Beach, FL, pp. 364–374, 1986.

[UnderKoffler 97]   J. Underkoffler. "A View From the Luminous Room," *Personal Technologies*, Springer-Verlag London Ltd., Vol. 1, pp. 49–59, 1997.

# Augmented Realities Integrating User and Physical Models

Thad Starner, Bernt Schiele, Bradley J. Rhodes, Tony Jebara, Nuria Oliver, Joshua Weaver, and Alex Pentland

**Abstract.** *The obvious advantage of wearable computing is mobility; it also offers the user a certain intimacy with augmented realities. A model of the user is as important as a model of the physical world for creating a seamless, unobtrusive interface while avoiding "information overload." This paper summarizes some of the current projects at the MIT Media Laboratory that explore user and physical environment modeling.*

## 1  Introduction

Wearable computers have the potential to *see* as the user sees, *hear* as the user hears, and experience the life of the user from his or her perspective. They are also physically and mentally more intimate with the user than are other kinds of computers and they may be used for extended periods of time, allowing a unique opportunity to model users' usage patterns and habits. This increase in user and environmental information can lead to more intelligent and fluid interfaces that incorporate the physical world.

This paper summarizes several of the current wearable computing and augmented reality research projects at the MIT Media Laboratory that explore the dimensions of user and physical modeling. (See Figure 1.) For more complete information on a particular project, the reader is encouraged to refer to the original papers.

## 2  Remembrance Agent

The Remembrance Agent (RA) is a program that continuously watches over the shoulder of the user of a wearable computer; it displays one-line summaries of notes files, old e-mail, papers, and other text information that might be relevant to the user's current context [Rhodes 97]. These

73

*Starner et al.*

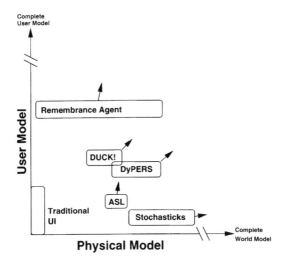

**Figure 1.** Wearable computing projects positioned along conceptual axes of the degree of involvement of a user model versus the extent to which the project models the physical environment of the user. The arrows indicate the direction of current research.

summaries are listed on a head-up display (HUD), so that the wearer can view the information with a quick glance. The full text can be retrieved using a one-handed keyboard.

The wearable version of the RA uses physical sensors to model the user's environment. The RA continuously monitors these sensors, as well as the notes being entered by the user. It uses this information to suggest documents, from a set of pre-indexed text, that are most relevant to the user's current situation. For example, a user's context might be described by a combination of the current time of day and day of the week (provided by the wearable system's clock), location (provided by an infrared beacon in the room), the person to whom the user is speaking (provided by an active badge), and the subject of the conversation (as indicated by the notes being taken). The suggestions provided on the RA are based on a combination of all of these elements, using text-retrieval techniques similar to those used in Web search engines.

## 2.1  Augmented Reality Remembrance Agent

The wearable RA uses an overlay display, but does not register its annotations with specific objects or locations in the real world — as one might expect from a full AR system. In many cases, such a real-world fixed dis-

**Figure 2.** Multiple graphical overlays aligned through visual tag tracking. The color code of each tag provides a unique ID for the object. In addition, the vision process tracks each tag in 2.5D (position, tilt, and distance from the camera) as the head-mounted camera moves.

play wouldn't even make sense, since suggestions are often conceptually relevant to the current situation without being relevant to a specific object or location. In order to examine AR interfaces, a different version of the RA was implemented using a desktop computer, HUD, and head-mounted camera. The wearer of the system viewed the world through the camera, and the camera output was transmitted wirelessly to a Silicon Graphics workstation. Around the room were colored tags, which a vision system could detect. The size and shape of the tags were used to determine their distance and orientation (see Figure 2). Color-coding was used to identify each tag.

Whenever a tag was identified, its unique code was referenced in a table, and the associated pre-specified message was overlayed on top of the object being viewed. The system could detect the distance of the tag, allowing more information to be displayed as the user came closer to a tagged object. Thus, the act of approaching a tag was equivalent to clicking on a hypertext link.

On top of this system, the Remembrance Agent showed the top suggestions for an individual tag. This created a two-level information system, with some information being provided by the infrastructure (tied to the tag via a lookup table), and some information provided from the user's own files via the Remembrance Agent.

## 2.2 Dynamic Personal Enhanced Reality Agent

A recent extension of the system described above uses a generic object recognizer to identify objects instead of tags. The system, called *Dynamic*

**Figure 3.** A DyPERS user listening to a guide during a test art gallery tour

*Personal Enhanced Reality System* (DyPERS [Schiele 99]), retrieves audio
and video clips based on associations with real objects. The generic object
recognizer is based on a probabilistic recognition system [Schiele 96], which
is capable of discriminating more than 100 objects in the presence of major
occlusions, scalings, and rotations. While 100 objects is not enough to be
practical in an unconstrained environment, the number of possible objects
can be significantly reduced using the location of the user, time of day, and
other available information. (See Figure 3.)

## 3   User-observing Wearable Cameras

In the previous section, head-mounted camera systems face forward and
observe the same region as the user's own eyes. By changing the angle of
the camera so that it points downward, we can use it to track the user's
own body. This allows the user's hands, feet, torso, and lips to be observed
without the gloves or body suits associated with virtual reality gear.

Two projects currently use this camera orientation. The first attempts
to translate American Sign Language into English by tracking the user's
hands via the downward-looking camera. The wearable ASL recognition
system outperforms equivalent desk-based camera systems and most
dataglove-based systems in recognition accuracy. For five-word sentences
composed from a 40-word lexicon, the system achieves 96.8% word accuracy with an unrestricted grammar (any word is possible, any number of
times, in any order) [Starner 98b].

The second system, DUCK!, begins to demonstrate how such methods
may be useful for augmented realities. Using only two video views, DUCK!
tracks the wearer's current location and task. DUCK!'s domain is limited to the real-space game Patrol, which is played by MIT students every
weekend in a campus building. Participants are divided into teams and

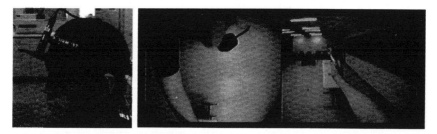

**Figure 4.** Left: The Patrol cap with two cameras. The larger, visible camera is mounted facing downward. The second camera faces forward and is hidden by the brim. Right: Images from the Patrol cap. The left and right images are from the downward-looking and forward-looking cameras, respectively.

aggressively hunt each other with rubber suction-cup dart guns through 14 rooms or areas. DUCK! monitors the average color and luminance values of the floor, the scene in front of the player, and the player's nose as a lighting calibration image. These images are pulled from the two video streams at six frames per second in order to determine location. In 24.5 minutes of training video and 19.3 minutes of test video, the system determines room location with 82% accuracy [Starner 98a]. (See Figure 4.) DUCK! also attempts to discriminate between the player's tasks by identifying hand gestures representing aiming/shooting, reloading, and everything else, through a combination of a generic object recognition system [Schiele 96] and the hidden Markov models (HMMs) used in the ASL task above. Preliminary results show 86% accuracy in distinguishing these classes. Other user actions, such as standing, walking, running, and scanning the environment, can be considered as tasks that can run concurrently with other actions. Future work will address these tasks.

While still in preliminary stages, the systems described above suggest how context perception may be used in augmented reality interfaces. Through heads-up displays, the players can keep track of the team's positions, even to the extent of "seeing through walls" to what may be occurring several rooms away. If aim and reload gestures are recognized by a particular player's system, his position can be highlighted in the rest of the team's displays, indicating that he needs aid. Furthermore, when the computer recognizes a player to be in battle, the computer should inhibit his interface in order to avoid interruptions.

## 4 Stochasticks

Stochasticks is a practical application of wearable computing and augmented reality which enhances the game of billiards [Jebara 97]. (See

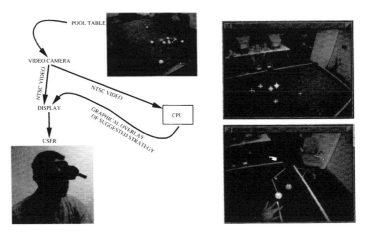

**Figure 5.** Left: The system components. Right Top: Finding the balls. Right Bottom: Suggested shot.

Figure 5.) Probabilistic color models and symmetry operations are used to localize the table, pockets, and balls through a video camera near the user's eyes. The system classifies the objects of interest and ranks each possible shot in order to determine its relative usefulness. The system allows the user to proceed through a regular pool game while it automatically determines strategic shots. The resulting trajectories are rendered as graphical overlays on a head-mounted live video display. The wearable video output and the computer vision system provide an integration of real and virtual environments which enhances the experience of playing and learning the game of billiards, without encumbering the player.

## 4.1   The System

A wearable computer is the hardware platform for the system and it includes a head-mounted display, a head-mounted video camera, and a central processing unit. The head-mounted camera is a miniature ELMO 2.2-mm video camera mounted on the heads-up display and aligned with the orientation of the eyes. Thus, the user's head direction will automatically direct the camera to areas of interest in the scene. The head-mounted display consists of a Virtual I/O 3D display (or a Sony Glasstron), which allows the CPU to project semi-transparent imagery into each eye via two separate CRTs at about 10 Hz.

Once the ball position is known, the easiest possible shot for a given player is computed, and the shot trajectory is projected onto the user's eye. At this point, we are undertaking a performance analysis of the overall system. The reliability of the algorithm is being investigated as well as its accuracy for both 2D and 3D overlays.

# 5 Conclusion

We have demonstrated how computer vision can be incorporated into augmented realities. We have discussed self-observing wearable camera systems that identify the user's gestures and location in a variety of conditions. Finally, through the projects presented, we have shown how modeling of both the user and the physical world play an important role in augmented realities.

# References

[Jebara 97]   T. Jebara, C. Eyster, J. Weaver, T. Starner, and A. Pentland. "Stochasticks: Augmenting the Billards Experience with Probabilistic Vision and Wearable Computers," *International Symposium on Wearable Computers*, IEEE Press, Cambridge, MA, pp. 138–145, 1997.

[Rhodes 97]   B. Rhodes. "The Wearable Remembrance Agent: A System for Augmenting Memory," *Personal Technologies, Special Issue on Wearable Computing*, Vol. 1, No. 1, Springer-Verlag, London, pp. 218–224, 1997.

[Schiele 96]   B. Schiele and J Crowley. "Probabilistic Object Recognition Using Multidimensional Receptive Field Histograms," *International Conference on Pattern Recognition*, Vol. B, Vienna, Austria, IEEE Press, pp. 50–54, August 1996.

[Schiele 99]   B. Schiele, N. Oliver, T. Jebara, and A. Pentland. "An Interactive Computer Vision System, DyPERS: Dynamic and Personal Enhanced Reality System," *International Conference on Computer Vision Systems*, Gran Canaria, Spain, Springer, pp. 51–65, 1999.

[Starner 98a]   T. Starner, B. Schiele, and A. Pentland. "Visual Conextual Awareness in Wearable Computing," *Second International Symposium on Wearable Computers*, Pittsburgh, PA, IEEE Press, pp. 50–57, 1998.

[Starner 98b]   T. Starner, J. Weaver, and A. Pentland. "Real-time American Sign Language Recognition Using Desk and Wearable Computer-based Video," *IEEE Transactions on Pattern Analysis and Machince Intelligence*, Vol. 20, No. 12, pp. 1371–1375, December 1998.

# Designing Interactive Paper: Lessons from Three Augmented Reality Projects

Wendy E. Mackay and Anne-Laure Fayard

**Abstract.** *Augmented reality (AR) offers a new paradigm for interacting with computers, linking familiar, physical objects to powerful computer networks. Although still in its infancy, the field is expanding rapidly, due to a wealth of new materials and technologies and a shift away from conventional notions about human-computer interaction. The increasing power and decreasing size of computers as well as the development of new materials such as electronic paper and ink make AR the user interface of the future.*

*However, creating successful AR applications is not simply a matter of coming up with a new technology, no matter how clever. Because of the potential for confusion between the real and the virtual, AR requires a multi-disciplinary approach, with a strong emphasis on the user. New technologies may offer fundamentally new ways of empowering users, but successful applications will have to be integrated into real-world activities.*

*We are interested in a particular type of augmented reality, which we call "Interactive Paper." We describe three projects: Ariel provides augmented engineering drawings for construction engineers; Video Mosaic provides augmented storyboards for video producers; and Caméléon provides augmented flight strips for air traffic controllers. We then examine paper from three perspectives: as a physical object, as a social artifact, and as an augmented object. We argue that field observations of users are essential for understanding the situated nature of the use of paper and that prototyping and participatory design are necessary for developing effective interactive paper applications.*

## 1 Introduction

Our work in augmented reality grows out of our observations of users who have held on to paper, despite powerful incentives to adopt electronic replacements. Contrary to what many believe, users are not Luddites, clinging to paper as a way of resisting change. On the contrary: most are excited by the benefits offered by computers and some are even accomplished hackers. Their resistance is, in fact, extremely practical. New computer systems are often less efficient or cannot perform required tasks. Depending upon the situation, users either reject the new system entirely (as do air traffic controllers who refuse to use on-screen flight strips) or else increase their workload by juggling an awkward combination of the existing paper and new on-line systems (as in many office applications).

Augmented reality offers an interesting solution, providing a direct link between physical paper and a computer network. This "interactive paper" has the potential to offer the best aspects of both paper and electronic documents, turning paper into the user interface. However, interactive paper is not a panacea. A lack of understanding of the differences between physical and electronic paper, particularly as they come into play in the context of the users' work, can result in serious confusion. How do we provide users with feedback when aspects of the augmented paper break down or get out of sync? What happens when there are discrepancies between the real and virtual forms of paper?

We have explored the concept of interactive paper, defined as using physical paper as the interface to a computer, in three different applications: Ariel provides augmented engineering drawings for construction engineers; Video Mosaic provides augmented storyboards for video producers; and Caméléon provides augmented flight strips for air traffic controllers. This paper briefly describes each application and then discusses interacting with paper from three perspectives: seeing it as a physical object, as a social artifact, and as an augmented object.

## 2    Ariel: Augmented Engineering Drawings

Our first interactive paper project, called Ariel [Mackay 93] [Mackay 95], grew out of work for the EuroCODE European ESPRIT project. We were initially charged with developing a multimedia communication system for construction engineers building the (then) longest suspension bridge in the world, in Denmark. In our observations and interviews with the engineers, we discovered that, although they all had computers in their offices, they rarely used them. Engineers were constantly on the move, from the boat, to the bridge, to pre-fabrication sites, to meeting rooms, and to their offices, making an office-based multimedia communication system impractical. We observed that engineers centered their work around their engineering drawings. Although in theory they had access to thousands of on-line drawings, in practice they used four or five at a time, which they folded and carried with them. Drawings were annotated by hand to reflect major and minor changes, a third of which were never entered into the computer. Thus the paper versions of the drawings contained a much more accurate description of the bridge that was actually built than did the electronic version.

We explored a variety of hardware and software configurations, based on discussions and collaborative prototyping sessions with the construction engineers, and decided to turn the drawings themselves into the interface to the multimedia communication system. The first prototype used a large (A0) size graphics tablet (designed for making maps) to capture commands and gestures made by the user (Figure 1a). Specialized software (written in C in Unix) projected relevant information onto particular drawings, letting users create, move, and send multimedia annotations and link them to the EuroPARC mediaspace [Mackay 92]. To calibrate, we projected an "X" on to each successive corner for the user to point at. Figure 1b shows a later (Macintosh) desktop version with a smaller (A2) graphics tablet, an Optima LCD video projector, a tiny video camera to detect a red light source (LED), and a hand-held scanner to capture hand-written annotations from the drawing. We experimented with turning blank sheets of paper (with an LED in the corner) into virtual windows, projecting menu and other information in the appropriate location on the desktop.

## 3    Video Mosaic: Augmented Storyboards

Our next interactive paper project was Video Mosaic [Mackay 94], designed to augment paper storyboards for video producers. This project grew out of the first author's collaborations with professional video producers, who rely on a paper artifact, the storyboard, in order to design and communi-

**Figure 1.** Figure 1a (left): Unix version of Ariel with A0 graphics tablet. Figure 1b (right): Light-weight desktop version of Ariel (Macintosh).

cate story ideas. Even those with access to powerful, multimedia computers continue to use paper storyboards to work out and share ideas. Some storyboards are formal, such as the pre-printed, translucent versions we used to create interactive multimedia software at Digital Equipment Corporation in the early 1980s. Others are simply sequences of sketches with annotations. Yet all storyboards share certain basic characteristics: providing a linear presentation of a story, using a series of sketches or images (the "best frame" of a clip), and including the script, shot information, and other annotations. Video producers often rearrange the elements of a storyboard, manipulating paper in three-dimensional space to organize the linear presentation of the story over time.

Video Mosaic lets video producers link paper storyboards to an online editing system, based on EVA [Mackay 89]. The first prototype used a video camera to capture commands from "paper buttons" and identify storyboard elements (Figure 2a). Tapping on the table (detected by a microphone) caused the camera to grab the current image, which was analyzed with optical character recognition to identify storyboard elements and user commands (play, stop, etc.). A small video screen embedded into the desktop displayed the resulting video, with related storyboard information projected onto the table next to it. Later prototypes placed control commands directly onto the paper storyboards (Figure 2b), making it easier to arrange individual elements and view the overall storyboard (Figure 2c). We also used barcodes to identify storyboard elements, eliminating the fixed camera.

## 4   Caméléon: Augmented Flight Strips

Our most recent project is Caméléon [Mackay 98], designed to augment paper flight strips for air traffic controllers. Despite massive efforts to replace flight strips, air traffic controllers around the world continue to rely on them to control air traffic. After spending four months with a team of en-route controllers in Paris, we discovered that paper flight strips serve a variety of functions that are difficult or impossible to duplicate with conventional computer interfaces. Over the next year, we worked with controllers to develop prototypes that could *track* the position of individual paper strips, *capture* information from them, and *display* information onto them. We were able to use pre-existing technologies for capturing information (video camera, graphics tablet, touch screen) and displaying information (video projection, regular monitor, touch screen), but had to develop our own technology for tracking. One solution used a frame with spring-loaded

**Figure 2.** Unix Video Mosaic with paper buttons, projected and embedded video.

**Figure 3.** Macintosh storyboard element with barcodes and control commands.

**Figure 4.** Macintosh version with barcodes and projected video.

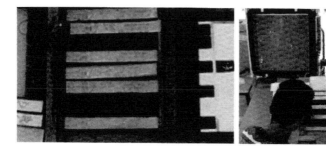

**Figure 5.** Figure 3a (left): Individual stripholders are detected by the electronic stripboard. Figure 3b (right): Caméléon captures annotations from a graphics tablet below the stripboard and presents information on an adjacent touchscreen or on the RADAR.

metal contacts along the sides (Figure 3a) to hold modified stripholders, each with an embedded resistor. Inserting a stripholder completed the circuit, identifying the precise location of that strip. When we combined this frame with a graphics tablet next to a touch screen (Figure 3b), we could capture all hand-written annotations and present information specific to each strip. We created an Interaction Browser to let controllers associate marks they made on individual strips with actions on the RADAR and on the training simulator.

## 5    Interacting with Paper

Our experiences led us to reflect upon all the physical characteristics of paper and its social role relate to the design of interactive paper applications.

### 5.1    Paper is a Physical Object

Paper has physical characteristics, called affordances, that affect the way people use it. Not only is it lightweight and flexible, but it is easy to annotate and personalize. People take advantage of minute, seemingly irrelevant details (a dog-eared corner, a coffee stain, a hand-written mark) to identify particular paper documents quickly. People go beyond officially-sanctioned uses, inventing new objects based on the situation at hand. A newspaper can be a child's hat, a paper airplane, or the lining of a bird cage; an envelope can be used as a scratch pad or a book mark. For video producers, storyboards are a creative outlet, and each person has an individual style. In contrast, bridge engineers work with other people's drawings and controllers work with pre-printed strips, annotating them in their own ways. Yet in all three cases, it is the users who decide on the manner in which they individually and collectively interact with these paper documents.

Our users were reluctant to give up paper, not because new computer systems did not provide reasonable imitations of the "official" uses of the paper, but because the new systems were completely unable to handle their invented uses of paper. Air traffic controllers reported that the physical act of writing helped them remember what they wrote, whereas a selection from a menu was rapidly forgotten. Holding a strip in one's hand is an easy and effective reminder of which plane one must handle next. Video producers also took advantage of the physical nature of paper, arranging the elements of a multimedia project's storyboard on the floor and walking down different paths to experience different story lines. Construction engineers walking around a site would scribble notes and identify problems by writing directly on the drawings they carried. Users in all three applications found paper to be fast and intuitive, whereas on-line versions were cumbersome and slow.

## 5.2 Paper is a Social Artifact

People use paper artifacts within a historical and social context. In all three applications, paper served as lightweight support for collaborative work. People working together have a peripheral awareness of each other, permitting low-cost, non-intrusive tracking of each other's work. For example, a controller could choose to place a new strip in a neutral area next to a colleague or lean over and physically insert it into the stripboard, indicating that the plane required immediate attention, without saying a word. Similarly, handing a physical design drawing to another engineer implied a hand-off of responsibility that simply did not work with less direct methods such as e-mail.

People who work together closely evolve certain conventions for interacting with their documents. The most sophisticated interactions were created by the controllers, who constantly refined their codes for marking flight strips via an on-going social process, responding to continued changes in air traffic, the air space, and air traffic regulations. Conventional computer systems rarely provide users with this level of flexibility or the ability to adapt the interface in the face of constant change.

## 5.3 Paper as an Augmented Object

Augmenting (rather than replacing) paper is an intriguing solution for applications in which the physical and social roles of paper are important.

However, interactive paper can also cause confusion. People intuitively understand the laws of physics with respect to real objects. People also learn how to use particular software applications. But what happens when these conflict in an augmented paper object? For example, we understand how to erase pencil from a paper document. We may also know how a particular text editor erases characters. But what happens when an electronic mark is erased and not its physical counterpart? Or vice versa? A key design consideration is thus the registration of the real and the virtual.

Note that this is not simply a technical question: providing feedback to users when the links are broken may be more important than ensuring that the links themselves are always there, particularly in safety-critical applications such as air traffic control. We were pleased when a controller complimented Caméléon by saying it was "invisible" and let him work as before while benefiting from new functionality. But it was also clear that we had to make sure that breakdowns were apparent and that controllers could work around them. We designed all our applications to support occasional or frequent, but never constant, connections between physical paper and on-line documents and ensured that the original uses of the paper document were always available.

We found that understanding how paper is used was important for deciding how best to augment it. For example, controllers write notes to themselves, which need not be captured; they communicate information with others, which must be shared but need not be interpreted; and they make decisions which must interpreted before being sent to software applications. Caméléon took advantage of these distinctions to perform only the on-line tasks that were necessary. Similarly, rearranging physical paper is easier than its on-screen equivalent, whereas editing video is best performed by the computer. Video Mosaic was designed to let video producers work with paper when that was most convenient, and switch to on-line functions when the computer could handle it best. Finally, the bridge engineers found scribbling on drawings to be easier than any on-line equivalent, but required technology to talk to each other. Ariel was designed to let them work with their drawings, but use the computer network when they wanted to communicate with each other (via a video-based media space) and share drawing changes. In each case, we examined the existing structure and social context of the paper documents before deciding how best to augment them.

## 6   Conclusions

The "paperless office" is clearly a myth: paper continues to be an essential component of many complex, collaborative work settings. Augmented reality provides a powerful alternative to the "keep it or replace it" choices traditionally faced by system designers. Yet, just as there is an almost infinite variety of kinds of paper and uses of paper, interactive paper offers an equally wide range of possible implementations. Although we are interested in the technological aspects of interactive paper, our main concern is the most effective way to embed it into real-world settings. By understanding pre-existing uses of ordinary paper, we were able to create a variety of design solutions that make sense to their users.

We do not believe in an "ideal" interactive paper technology: minor technical differences in the ways information is captured, interpreted, and presented, as well as differences in modes of registering the real with the virtual, can cause major differences in the way the application is perceived and used. The situated nature of paper has important design implications: designers must consider how users will adopt and co-opt interactive paper in the context of their daily work. We urge augmented reality designers to explore the design space of possible solutions with users, rather than immediately seeking a single technical solution. The goal is to design augmented reality applications that fit smoothly into the real world.

# References

[Mackay 89]  W.E. Mackay, and G. Davenport. "Virtual Video Editing in Interactive Multi-Media Applications." *Communications of the ACM*, Vol. 32, No. 7, July 1989.

[Mackay 92]  W.E. Mackay, "Spontaneous interaction in virtual multimedia space: EuroPARC's RAVE system." *Imagina '92*, Monte Carlo, Monaco, 1992.

[Mackay 93]  W.E. Mackay, G. Velay, K. Carter, C. Ma, and D. Pagani. "Augmenting Reality: Adding Computational Dimensions to Paper." Eds. P. Wellner, W. Mackay, and R. Gold. *Computer-Augmented Environments: Back to the Real World.* Special issue of *Communications of the ACM*, Vol. 36, No. 7, 1993.

[Mackay 94]  W.E. Mackay, and D. Pagani. "Video Mosaic: Laying out time in a physical space." *Proceedings of Multimedia '94.* San Francisco, ACM, 1994.

[Mackay 95]  W.E. Mackay, D.S. Pagani, L. Faber, B. Inwood, P. Launiainen, L. Brenta, and V. Pouzol. "Ariel: Augmenting Paper Engineering Drawings." Videotape Presented at CHI '95, 1995.

[Mackay 98]  W. Mackay, A.L. Fayard, L. Frobert, and L. Médini. "Reinventing the Familiar: Exploring an Augmented Reality Design Space for Air Traffic Control." *Proceedings of ACM CHI '98 Human Factors in Computing Systems,* Los Angeles, ACM/SIGCHI, 1998.

# AREAS: Augmented Reality for Evaluating Assembly Sequences

Jose Molineros, Vijaimukund Raghavan, and Rajeev Sharma

**Abstract.** *Augmented reality provides a powerful and intuitive interface that can enhance the user's understanding of a scene. We consider the problem of scene augmentation in the context of the assembly of a mechanical object. Concepts from robot assembly planning are used to develop a systematic framework for presenting augmentation stimuli for this assembly domain. We then describe an interactive evaluation tool called AREAS, which uses augmentation schemes for visualizing and evaluating assembly sequences. The system also guides the user step–by–step through an assembly sequence.* Computer Vision, *together with a system of markers, provides the sensing mechanism necessary to interpret the assembly scene.*

## 1  Introduction

An augmented reality system that can enhance a user's view of the surrounding scene, with annotations based on the scene content, has many potential applications. These include: aiding in repair or maintenance [Feiner 93], manufacturing, education, training, medicine [Bajura 92], battlefield maneuvers, etc.

We consider the problem of scene augmentation in the context of assembling a mechanical object from its components. Concepts from robot assembly planning are then used to develop a systematic framework for presenting augmentation stimuli for this assembly domain. In this paper, we use augmentation to aid in assembly evaluation. Figure 1 conceptually illustrates an AR system that will be considered for our discussion. The goal of this system is to help evaluate the feasibility and efficiency of a particular sequence in the assembly of a mechanical object from its components. This is done by using augmentation to guide the operator through each step in the sequence. The user can prompt the system for detailed information at any time. The augmentation is provided with the

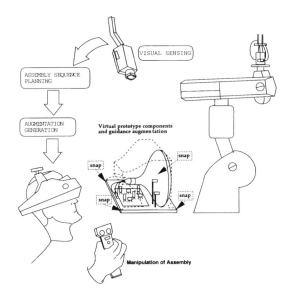

**Figure 1.** The AREAS system and its components.

help of 3-D graphics superimposed over the assembly scene. There could also be additional information displayed on a nearby computer monitor. The feedback about the current content of the scene is provided to the AR system with the help of a set of video cameras, together with the necessary computer vision algorithms [Sharma 97]. Vision is also used to record the actual moves of the user. This produces a convenient flow of information between the design engineer and a computer assembly planner.

Using the augmented reality interface, it would be possible to realize the concept of *mixed prototyping*, where part of the design is available in a physical prototype and part of it exists only in the virtual form. This can be an improvement over pure virtual prototyping [Gupta 97] in cases where more realistic feedback on the assembly is needed.

In this paper, we present the design of a proof-of-concept assembly evaluation tool we call AREAS (Augmented Reality for Evaluating Assembly Sequences) and explore its use for evaluating assembly sequences using the concept of mixed prototyping.

## 2   Background: The Assembly Domain

The assembly involves starting from a set of parts and putting them together in a proper sequence for creating an assembled object. A completed

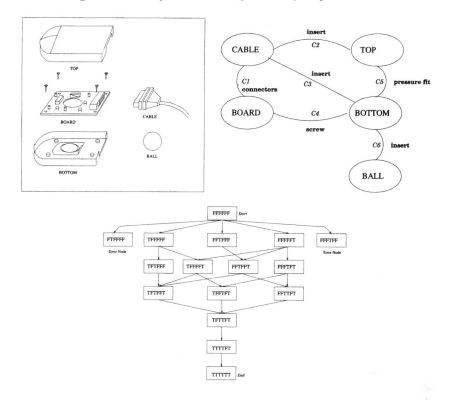

**Figure 2.** The "mouse" assembly, its liaison graph, and the assembly graph.

assembly is composed of *parts*, some of which are connected together to form *subassemblies*. The interconnections or liaisons between the parts of an assembly can be represented in the form a *liaison graph* (see Figure 2).

One way of representing the state of an assembly is in terms of an array of boolean variables for each of the liaisons [Homem de Mello 91]. In terms of the assembly states, the goal of the assembly task is to transform states from the initial node or start state $[FF \cdots F]$ to the goal state $[TT \cdots T]$. Each assembly operation results in a single change in the assembly state. Thus, all possible assembly transformations can be represented in terms of a directed graph that we term *assembly graph*. Error nodes are physically unreachable states or subassemblies with unstable conditions. Any path connecting the node $[FFFFF]$ to the node $[TTTTT]$ is a valid assembly sequence if it does not include an error node. This representation is adapted and used for monitoring the assembly operations, controlling the augmentation stimulus, and helping in error recovery.

# 3   The AREAS System

The key feature of the proposed system is the novel and intuitive way in which a design engineer interacts with the computer assembly planner. To handle the sequencing of large and complex assemblies, computer-based tools have been proposed [Baldwin 91]. These interact with the user mainly through a set of predefined questions. Instead of using this human-computer interface which requires the engineer to answer a series of questions (using some graphical aids) in order to help the assembly planner, the augmented reality interface would allow the user actually to manipulate the prototype assembly components and step through a computed assembly plan for the purpose of evaluation. This permits the engineer to detect whether a specified assembly operation is infeasible or difficult. At the same time, the engineer can edit the assembly plan based on a trial run of an assembly sequence. The visual sensing system automatically monitors and records the moves made by the user.

The critical flow of information between the user and the assembly planner is as follows:

- *To the user from the assembly planner.* The planner gives a computed assembly sequence to the user in the form of multimodal augmentation that guides the user through each of the needed assembly step and then queries about feasibility or cost of the operation.

- *To the assembly planner from the user.* As the user manually carries out the assembly steps (which may be different from the ones suggested by the planner), the actual sequence is recorded automatically. The user answers any explicit queries by the planner and gives any optional cost information. The assembly moves are also recorded in the form of video clips for later review.

The above interaction is enabled by the three components of the AR-EAS system: *assembly planning, multimodal augmentation generation,* and *scene and assembly state sensing.* Figure 1 shows a functional description of AREAS and the flow of information among its components. The main functional units shown participate in an interactive loop with the human operator.

## 3.1   Planning to Control Augmentation

The proposed approach for assembly sequence evaluation requires co-operation between the user and an automatic assembly sequence planner

[Ames 95]. This planner would take a representation of the design of a product whose assembly sequence has to be planned, and come up with the best assembly sequencing alternatives.

The planner compiles all the possible assembly sequences, and searches them using human input and interactive automated search, to identify the best assembly sequence. During this process of evaluation, the planner communicates with the other two modules of the AREAS system. It also communicates with the user through the augmentation. This enables the user to contribute to the evaluation and also to control the entire AREAS system.

The planner takes as input the liaison graph of the assembly whose sequence is to be evaluated, and then generates all the possible sequences based on the precedence constraints which are also given as input. This results in automatic pruning of the assembly graph, and reduces the search space for the interactive stage of the planning.

The planner then traverses paths in the assembly graph during the evaluation, guiding the user through the liaison (corresponding to a change in state in the assembly graph), using multimodal augmentation, and, at each node in the graph, accepting the user's evaluation of the liaison after verifying its completion using the visual tracker. The information the user gives the planner on the difficulty of completing a liaison is used cumulatively to compute the cost of the sequence under evaluation and to eliminate other sequences.

The state of the assembly determines what to display. Thus, a node of the assembly graph helps determine the corresponding augmentation display—for example, by displaying the choice of the next move. The planner then passes the current assembly status to the augmentation system. The augmentation system then determines the appropriate stimulus to be presented to the user based on the assembly state.

Any action by the human operator can be viewed in terms of the assembly graph. When the user connects two parts together, the state of the assembly changes, and this is represented by a path in the graph from one node to the next. Thus, the liaison and assembly graphs provide all the information required to evaluate the sequence. That permits the user to answer all the planner queries through graphical interfaces provided by the augmentation system. The information is then passed to the planner and converted into its internal representation.

Information also flows from the planner to the user. When the planner completes an evaluation step such as pruning the assembly choices in its representation, information is passed to the augmentation system so that the user knows which sequences are valid.

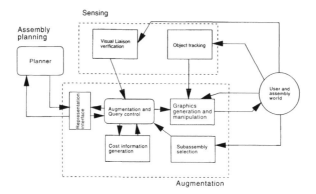

**Figure 3.** Information flow within AREAS components.

## 3.2   Augmentation in AREAS

We have implemented a tool for developing, visualizing, and evaluating
AR presentation schemes for the assembly domain; this tool functions as
the main augmentation generation engine of the AREAS system.   The
tool is called AUDIT (AUgmentation Development Interactive Tool), and
it provides the augmentation block seen in Figures 1 and 3. A detailed
description can be found in [Sharma 96].

The four main functional components involved in AUDIT are: (1) augmentation and user query handling: (2) assembly visualization, augmentation display, and virtual object manipulation; (3)   liaison cost assignment:
and (4) virtual subassembly selection for visualization.

Figure 3 is an overview of the augmentation engine, as well as its interaction with the rest of the AREAS system. The main functional units are
shown along with the flow of information between them. In order to have
a modular design, all the components function in independent windows.

User query handling operates in the decision space window (DSW). The
DSW answers queries from the user about the present state of the assembly
and the evaluation, as well as providing information on how to complete
the next possible assembly liaisons. The user can ask the system how to
perform a liaison by clicking on one of the nodes of the assembly graph
(see [Sharma 96]). The DSW will provide augmentation stimuli to the user
for guidance. It will show the user the completion of the liaison through
computer graphics as well as video and audio sequences. The DSW lets the
user see the assembly path followed and also error nodes (assembly states
that are forbidden or unreachable).

The user inputs the costs of each liaison through the DSW. When the
user clicks on any liaison on a displayed assembly graph, a cost input win-

dow will pop up, and the user can input the relevant cost information. The DSW uses different kinds of stimuli to aid in guidance. These stimuli are are displayed with the help of the graphics generation module and include simple labels and graphic primitives, 3–D models of parts, sound, and voice and video sequences.

The interface to the augmentation generation engine AUDIT is called the manipulation window. This window handles the graphics generation and manipulation. The DSW communicates what graphics to display, and the augmentation generation displays them according to the situation. The objective of this window is two-fold. First, the window permits the user to specify the level of detail of the interactive information presented to him. This information includes situated graphic primitives (see [Sharma 96]), audio sequences, and 3–D CAD models. The level of augmentation necessary to guide the user though an assembly decreases with his expertise in assembling that particular sequence. For an unexperienced user, a large amount of information is useful. For an experienced one, the cluttered display will only interfere with his effective completion of the task.

The second objective of this window is to test and evaluate the augmentation scheme and its usefulness. This is done by visualizing the scheme for completing a simulated assembly sequence (or sub-sequence) using graphic models of the assembly parts instead of real parts. [Sharma 96] presents examples of simulated sequences and augmentation aids, such as arrows, connecting lines, graphic primitives, and sounds. The user can check the usefulness of the augmentation scheme in the simulation.

The subassembly selection module implements an interface in a separate window that permits the user to select a part or subassembly for visualization. Finally, in the current implementation, the augmentation graphics are displayed in a monitor over live video in a window called the video window. This window is where the user can see the augmentation for assembly guidance and 3–D virtual parts for mixed prototyping. The graphics display is currently being implemented in a see-through HMD. Figure 4 shows two parts of a tool vise assembly: one real, and one virtual, as seen through the HMD. The information provided by tracking real parts in the assembly using computer vision updates the graphics pose. Issues involved in the practical implementation of an augmented reality interface can be found in [Azuma 97].

## 3.3   Monitoring Assembly with Vision

The assembly domain permits the use of specially designed markers to simplify the computer-vision algorithms and make real–time implementation

**Figure 4.** 3-D CAD models combined with real parts as seen through the HMD.
The parts are tracked using computer vision and a marker scheme.

feasible. A single video camera is currently used to capture the image of
the relevant portion of the assembly scene (more cameras can be used),
and a Minimum Spanning Tree-based coding scheme is used to process the
image for extracting the markers. These are then used to recognize the
individual parts, their positions and liaisons, and hence the state of the
assembly. This information is then passed to the augmentation module to
display the augmentation in an HMD and an adjacent monitor (or monitors). A comprehensive description of the system and algorithms can be
found in [Sharma 97] [DeMenthon 92].

## 4   Conclusions

This paper proposes an interactive, augmented reality tool that uses computer vision for assembly sequence evaluation and guiding. This is possible
because of a marker-based computer-vision technique and the use of an
augmentation scheme tailored for assembly. When applied to the assembly domain, computer vision helps in breaking the assembly domain down
into states and in interpreting the actions of the user. The resulting representation of the assembly permits the design of a guiding scheme using
augmentation stimuli.

## References

[Ames 95]                    A.L. Ames, T.L. Calton, R.E. Jones, S.G. Kaufman, C.A.
                             Laguna, and R.H. Wilson. "Lessons Learned from a Sec-
                             ond Generation Assembly Planning System," *IEEE In-
                             ternational Symposium on Assembly and Task Planning*,
                             IEEE Press, pp. 41–47, 1995.

[Azuma 97]       R. Azuma. "A Survey of Augmented Reality," *Presence: Teleoperators and Virtual Environments*, Vol. 6, No. 4, MIT Press, pp. 355–385, August 1997.

[Bajura 92]      M. Bajura, H. Fuchs, and R. Ohbuchi. "Merging Virtual Objects with the Real World: Seeing Ultrasound Imagery Within the Patient," *Computer Graphics*, Vol. 26, No. 2, pp. 203–210, 1992.

[Baldwin 91]     D.F. Baldwin, T.E. Abell, M.M. Lui, T.L. DeFazio, and D.E. Whitney. "An Integrated Computer Aid for Generating and Evaluation Assembly Sequences for Mechanical Products," *IEEE Journal of Robotics and Automation*, Vol 7, No. 1, pp. 78–94, 1991.

[DeMenthon 92]   D. DeMenthon and L.S. Davis. "Exact and Approximate Solutions of the Perspective-three-point Problem," *IEEE Transactions on Pattern Analysis and Machine Intelligence*, Vol. 14, No. 11, pp. 1100–1105, 1992.

[Feiner 93]      S. Feiner, B. MacIntyre, and D. Seligmann. "Knowledge-based Augmented Reality," *Communications of the ACM*, Vol. 36, No. 7, pp. 52–63, 1993.

[Gupta 97]       R. Gupta, T. Sheridan, and D. Whitney. "Experiments Using Multimodal Virtual Environments in Design for Assembly Analysis," *Presence: Teleoperators and Virtual Environments*, Vol. 6, No. 3, pp. 318–338, 1997.

[Homem de Mello 91] L.S. Homem de Mello and A.C. Sanderson. "Representations of Mechanical Assembly Sequences," *IEEE Journal of Robotics and Automation*, Vol. 7, No. 2, pp. 211–227, 1991.

[Sharma 96]      R. Sharma and J. Molineros. "Interactive Visualization and Augmentation of Mechanical Assembly Sequences," *Graphics Interface '96*, pp. 230–237, May 1996.

[Sharma 97]      R. Sharma and J. Molineros. "Computer Vision-based Augmented Reality for Guiding Manual Assembly." *Presence: Teleoperators and Virtual Environments*, Vol. 6, No. 3, pp. 292–317, June 1997.

# Integrating Augmented Reality and Telepresence for Telerobotics in Hostile Environments

John Pretlove and Shaun Lawson

**Abstract.** *Tasks carried out remotely via a telerobotic system are typically complex, occur in hazardous environments, and require fine control of the robot's movements. Telepresence systems provide the teleoperator with a feeling of being physically present at the remote site. Stereoscopic telepresence has been successfully applied to increase the operator's perception of depth in the remote scene, and this sense of presence can be further enhanced using computer-generated stereo-graphics to augment the visual information presented to the operator.*

*The Mechatronic Systems and Robotics Research Group (MSRR) has, since 1988, been developing high-performance, active stereo-vision systems culminating in the latest — a four degree-of-freedom, stereo-vision remote head. This system, used as part of a telepresence system, displays high-quality video to an operator wearing a head-mounted display or viewing a stereo, three-dimensional monitor. A sensor on the head-mounted display is used to provide the demand signal to drive the gaze point of the remote cameras.*

*This paper describes the Active Telepresence System and the development of an augmented reality (AR) interface. The results of preliminary experiments using the initial enhancements will be presented here, and we anticipate that the results from trials on a real application for the robotic inspection of sewers will be presented in the near future.*

## 1 Teleoperation and Telepresence

A teleoperator or a telerobot is a machine or device which extends an operator's sensing and manipulation capability to a remote environment. Today, modern and advanced teleoperators and telerobotic devices are finding numerous applications in hazardous environments, including: (1) those which

101

are hazardous to humans (such as nuclear decommissioning, inspection, and waste handling; bomb disposal and minefield clearance; unmanned underwater inspection; and search and rescue); (2) those in which humans adversely affect the environment (such as medical applications and cleanroom operations); and (3) those which are impossible for humans to be situated in (such as deep space and nano-robotics).

Telerobotics are finding applications in these areas because the technology can save lives and reduce costs by removing the human operators from the "theatre of operations." However, in most of these areas we still need humans in the control loop because of their very high level of skills and because machine technology is insufficiently advanced to operate autonomously and intelligently in such complex, unstructured, and often cluttered environments. This is a good example of Brooks's Intelligence Amplification, whereby machines, computers, and humans work in harmony to achieve what none could achieve alone [Rheingold 92].

To operate effectively in the remote environment, the operator requires sufficient visual information to be able to interpret the remote scene and undertake the task effectively and efficiently; this is usually accomplished by using a telepresence system. Such a system displays high-quality visual information about the task environment, and does so in such a natural way that the operator feels physically present at the remote site. Not only is it important to provide sufficient visual bandwidth, but it is also much more effective if the operator can control the positioning of the remote cameras. The MSRR group's Active Telepresence System achieves this by slaving the remote camera system to the motion of the operator's head. A generalized schematic of the MSRR's group Active Telepresence System is shown in Figure 1.

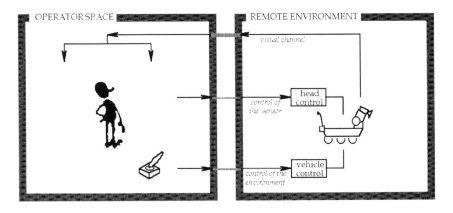

**Figure 1.** Generalized schematic of a teleoperator

Sheridan proposes that there are three principal and independent determinants of the sense of presence in a remote environment: the extent of sensory information (ideally the same level of sensory information that operators would have if they were physically placed in the remote environment), the control of the sensors (the ability to modify the position of the sensing device), and the ability to modify the remote environment (to be able to change objects in the remote environment or their relationship to one another) [Sheridan 92].

The success of this approach is due, in part, to visual capture (the tendency to believe the eyes above all other senses) and to the concept that humans reconstruct and understand their environment by active exploration rather than by viewing static scenes. It also allows the operator to concentrate fully on the task in progress rather than spend time manipulating the cameras in order to provide the best view.

The interface to telerobotic and telepresence systems has been extended by using augmented reality. A thorough survey of augmented reality is presented by Azuma [Azuma 97]. At a basic level, virtual reality modeling can be combined with the video stream to provide the user with additional information such as the robot position and orientation, heading, gaze angle, etc. This may be thought of as a virtual instrument panel.

At a higher level, the user might, for instance, position a three-dimensional stereo cursor on part of the real-world scene. Provided that the real and virtual worlds are calibrated and registered, the (x, y, z) position of the cursor in real-world coordinates would be available. This is a very powerful capability which would enable the robot or manipulator to move to that point; to measure the distance between two such points; and even to build or modify interactively a nominal three-dimensional computer model of the real world. In unstructured, remote environments, virtual objects from a toolbox could be attached to the real world. For instance, virtual barriers might be placed to protect regions that were out of bounds or dangerous; markers and labels could be left on objects for further processing; and virtual trajectories could even be defined for the robot or manipulator to follow. The MSRR group at the University of Surrey has been developing such systems for the remote inspection of hazardous environments for a number of years [Pretlove 94] [Pretlove 95] [Pretlove 96], and the following sections gives more specific details of this work.

## 2 Augmented Reality and Telepresence System

The Active Telepresence System consists of three distinct elements: the stereo head, its controller, and the display device. The stereo head, shown

**Figure 2.** Surrey Active Telepresence System

in Figure 2, is a flexible, light-weight structure which can be mounted onto a wide range of vehicles and other devices. It allows independent vergence and common pan and tilt of the twin CCD cameras, and provides dynamic, independent, and simultaneous control of all four degrees of freedom with dynamic characteristics similar to those of the human head. The system tracks the motion of the operator's head using a Polhemus or Intersense sensor mounted on the head-mounted display [Pretlove 96]. This results in a system whose use is intuitive, which has a latency of less than 75 ms, and which exhibits no appreciable overshoot.

The augmented reality system consists of an Intergraph PC workstation running WorldToolKit. The stereo graphics are combined with the video images of the remote scene using two scan converters, and these signals are then passed on to the head-mounted display and the stereo, three-dimensional monitor (see Figure 3). In order to present the illusion that the real and virtual objects coexist in the same space, the computer-generated images must be registered with the real scene. Accurate registration is important because of the ability of the human vision system to detect even small registration errors due to the high resolution of the fovea and the sensitivity of the visual system to differences. In some applications of AR and telepresence, the operator could tolerate small registration errors, and in most applications registration errors could have serious consequences. The method of calibration of the Surrey AR telepresence system is detailed by Wheeler [Wheeler 96].

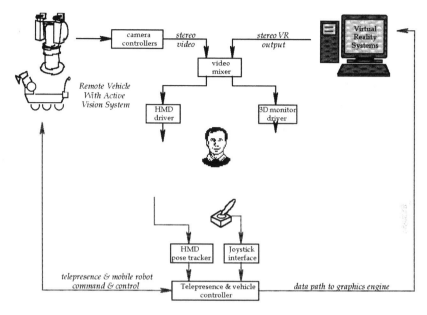

**Figure 3.** Schematic of the integrated Augmented Reality and Active Telepresence System

## 3 Experimental Work

We have carried out a number of experiments with the Augmented Reality and Active Telepresence Systems in a lab environment in order to demonstrate the benefits of providing additional task information to the operator (in the form of a three-dimensional virtual instrument); and we have implemented an interactive, three-dimensional cursor that can be used to interact with objects in the real world. For both experiments, the Active Telepresence System was mounted on a small, mobile lab robot (see Figure 3), and the results are presented in greater detail in Wheeler et al. [Wheeler 96]. Both experiments clearly demonstrate the advantage of using additional information in poor-visibility tasks and the clear advantages of being able to interact with a remote environment.

At the International Workshop on Augmented Reality (IWAR '98), we presented some up-to-date work on a new project to demonstrate the use of augmented reality as an aid for robotic pipe inspection and maintenance. This project is concerned with the examination of underground sewage pipes during robotic inspection and maintenance tasks. Wireframe or rendered graphics are overlayed on, and registered with, the two video streams as they are displayed to the human observer. This enables the position of joints, previous repair work, and previously detected defects in the pipe work to be displayed to the user. The images may also be fed to an image processing system for automatic analysis of pipe structure position and automatic defect detection. We plan to enable the position of the camera system to be relayed to a graphics workstation which contains a model of the underground pipe structure.

## 4    Conclusions

This paper describes the Active Telepresence System with an integrated augmented reality (AR) system designed to enhance the operator's sense of presence in hazardous environments. The Active Telepresence System has progressed a long way toward fulfilling the three components of presence outlined by Sheridan. The user sees the remote world in stereo vision through two cameras which are slaved to the motion of the user's head. This allows the operator to concentrate fully on the task in progress rather than spending time manipulating the cameras to provide the best view, and aids the visual search and the natural human habit of active exploration. The control of the cameras is instinctive, and, with a total system latency of the order of 75 ms, a high sense of presence is achieved. The user interface is enhanced by overlaying computer-generated graphics from a virtual reality system. Once input had been calibrated so that the real and the virtual were registered, a virtual instrument panel and an interactive, three-dimensional cursor were developed to demonstrate the potential benefits that such a system offers.

## References

[Azuma, 97]    R.T. Azuma. "A Survey of Augmented Reality." *Presence: Teleoperators and Virtual Environments* Vol. 6, No. 4, pp. 355–385, 1997.

[Pretlove 94]    J.R.G. Pretlove. "Stereo Vision: Crossing the Is and Dotting the Ts." *Industrial Robot.* Vol. 21, No. 2, pp. 24–27, 1994.

[Pretlove 95]    J.R.G. Pretlove and R.J. Asbery. "The Design of a High Performance Telepresence System Incorporating an Active Vision System for Enhanced Visual Perception of Remote Environments." *Telemanipulator and telepresence technologies II. Proceedings of SPIE.* Vol. 2590, ed. M. Salganicoff, pp. 95–106, 1995.

[Pretlove 96]    J.R.G. Pretlove. "Get a head in telepresence: Active vision for remote intervention." *Remote Techniques for Hazardous Environments,* ed. by D.W. James, British Nuclear Energy Society, London, pp. 264–269, 1996.

[Rheingold 92]    H. Rheingold. *Virtual Reality.* Mandarin, London, 1992.

[Sheridan 92]    T.B. Sheridan. *Telerobotics, Automation and Human Supervisory Control.* The MIT Press, London, 1992.

[Wheeler 96]    A.C. Wheeler, J.R.G. Pretlove, and G.A. Parker. "Augmented Reality to Enhance an Active Telepresence System." *Telemanipulator and Telepresence Technologies III, Proceedings of SPIE,* Vol. 2901, ed. by M.R. Stein, pp. 81–89, 1996.

# Part III

# Registration for Augmented Reality

*One of the major hurdles faced by augmented reality applications is the problem of precise registration of virtual elements with the real world. To date, it appears that the greatest potential for the highest precision comes from applying computer-vision-based registration techniques. In the following section, several approaches for using computer vision methods are described. The papers address such key issues as finding feature correspondences, speed, robustness, scalability, and multi-sensor fusion.*

# Computer Vision Methods for Registration: Mixing 3D Knowledge and 2D Correspondences for Accurate Image Composition

Gilles Simon, Vincent Lepetit, and Marie-Odile Berger

**Abstract.** *We focus in this paper on the problem of adding computer-generated objects to video sequences. A robust, two-stage statistical method is used for computing the pose from model-image correspondences of tracked curves. This method is able to give a correct estimate of the pose even when tracking errors occur. However, if we want to add virtual objects into a scene area which does not contain (or contains few) model features, the reprojection error is likely to be large. In order to improve the accuracy of the viewpoint, we use 2D keypoints that can be matched easily in two consecutive images. Since the relationship between two matched points is a function of the camera motion, the viewpoint can be improved by minimizing a cost function which encompasses the reprojection error as well as the matching error between two frames. The reliability of the system is demonstrated on an encrustation of a virtual car in a sequence of the Stanislas square.*

## 1 Introduction

Augmented reality (AR) is an effective means of exploiting the potential of computer-based information and databases. In AR, the computer provides additional information that enhances or augments the real world, rather than replacing it with a completely virtual environment. In contrast to virtual reality, where the user is immersed in a completely computer-generated world, AR allows the user to interact with the real world in a natural way. This explains why interest in AR has substantially increased over the past few years, and why so many medical, manufacturing, and urban planning applications have been developed [Berger 96] [Ertl 91] [Uenohara 96].

111

We focus in this paper on the problem of adding computer-generated objects (also called virtual objects) into video sequences. This is one of the key issues involved in numerous AR applications: for instance, suppose we want to assess the potential impact of a new construction in its final setting; visualizing the architectural project using a video of the environment allows designers to test several architectural projects on computer simulations alone. Special effects in movies also require such a composition process.

In order to make AR systems effective, the computer-generated objects and the real scene must be combined seamlessly so that the virtual objects align well with the real ones. It is therefore essential to determine accurately the location and the optical properties of the cameras. The registration task must be achieved with special care, because the human visual system is very good at detecting even small misregistrations. Realistic merging of virtual and real objects also requires that objects behave in a physically plausible manner in the environment: they can be occluded by objects in the scene, they are shadowed by other objects, etc.

In this paper, we focus only on the registration problem because it is one of the most basic challenges for augmented reality. But we have proposed some preliminary solutions to the occlusion problem in [Berger 97].

Registration problems can be solved by using either algorithmic solutions or sensor-based solutions. For instance, position sensors (such as Polhemus sensors) can be used to locate the camera (or the viewer) [State 96]. Easily detectable landmarks placed in the scene can also be used to make the registration process easier [Ertl 91]. However, instrumenting the real world is not always possible, especially for vast or outdoor environments. Vision-based object registration is an interesting and cheaper approach that leaves the environment unmodified.

In our approach, we assume that the camera has been calibrated previously. This means that the internal characteristics of the camera (focal length and optical center) have been previously computed using a calibration target [Faugeras 86]. Thus, the registration problem amounts to computing the location (or the viewpoint) of the camera.

Pose recovery has been extensively studied in the past few years. Two broad classes of methods can be distinguished: the classical one uses object-based registration. This means that 3D knowledge is needed to compute the pose from image/model correspondences. Such methods minimize the reprojection error of the model's features in the image. Hence, one of the limitations of this method originates in the spatial distribution of the model points: the reprojection error is likely to be large far from the 3D features used for the viewpoint computation.

The other alternative is basically 2D: if the projection of a sufficient number of points is observed from different positions, the camera pose can be recovered as well as the 3D location of the points up to a scale factor [Faugeras 86] [Tomasi 92]. Unfortunately, these approaches turn out to be very sensitive to inaccuracies in 2D feature measurements.

Our approach combines the strengths of these two methods: the classical viewpoint computation from 3D-2D correspondences is used to compute a first approximation of the viewpoint. Then, automatic extracting and matching of key-points between two consecutive views allow us to refine the viewpoint computation. As a result, viewpoint computation can be achieved through the sequence with minimal 3D knowledge on the scene.

Our system stands out from previous works on the following points:

- **Minimal interaction with the user**: the user is only asked to point out four points and their 3D counterparts in the first image of the sequence. The 2D features corresponding to the visible model features are then automatically determined and tracked in the sequence (the tracking tool we use is described in [Berger 94]). The pose is then computed from their correspondences with the 3D model.

- **Model features**: Since the world around us is not piecewise-planar, the significant features that can be extracted in an image are often curved contours, especially if we consider outdoor environments. One of the strengths of our pose algorithm is its ability to handle point, line, and curve correspondences for pose computation.

- **Robustness**: It is well known that false 3D-2D matches (outliers) can have a significant effect on the resulting pose. We have therefore developed a robust statistical method for computing the pose from the matching that results from the tracking process. Our approach has some points in common with [Ravela 96]. Ravela et al. also use robust methods for pose computation after the tracking stage. But they only consider point features, whereas we consider points, lines, and curved features.

- **Accounting for interest-point correspondences** Mixing 3D-2D correspondences in an image with 2D-2D correspondences between two consecutive frames allows us to compute the viewpoint accurately. Unlike model-based methods which require knowledge of a large number of 3D features in order to give precise results, our mixing method requires little 3D knowledge of the scene.

We begin with an explanation of our robust algorithm for viewpoint computation from 3D-2D correspondences, including results on various augmented reality applications. Section 3 describes how the viewpoint computation can be dramatically improved using key-point correspondences. Finally, we show results demonstrating the accuracy of pose estimation.

## 2 Robust Pose Computation from Various Features

Once the model/image correspondences have been established in the first frame, they are generally maintained during tracking. Unfortunately, tracking errors will sometimes result in a model feature's being matched to an erroneous image feature. Even a single such outlier can have a significant effect on the resulting pose. Robust approaches allow point features to be categorized as outliers or non-outliers [Haralick 89]. When curved features are considered, the problem is not so simple, since some parts of the 2D curves might match perfectly the 3D model while others might be erroneously matched. While numerous papers are dedicated to pose estimation from points or lines [Dementhon 95] [Ferri 93] [Kumar 94], only a few have been devoted to the 3D-2D registration of curves [Feldmar 97] [Kriegman 90]. However, even these papers are not concerned with possible matching errors. The details of our robust pose computation algorithm (RPC) are given in this section. Emphasis is put on the robustness of the computation. First we address the problem of point correspondences; the much more difficult case of curve correspondences is then considered.

### 2.1 Pose Computation from Point Correspondences

The problem consists in finding the rotation $\mathbf{R}$ and the translation $\mathbf{t}$ which map the world-coordinate system to the camera-coordinate system. Therefore, if the intrinsic parameters of the camera are known (they can be determined by a calibration process [Faugeras 86]), we have to determine six parameters (three for $\mathbf{R}$ and three for $\mathbf{t}$), denoted by vector $\mathbf{p}$.

We suppose we know the 3D points $M_{i\{1 \leq i \leq n\}}$ and their corresponding points $m_{i\{1 \leq i \leq n\}}$ in the image. Computing the viewpoint amounts to finding the $\mathbf{R}, \mathbf{t}$ which minimize the re-projection error:

$$\sum_{i=1}^{n} r_i^2 = \sum_{i=1}^{n} Dist(m_i, Proj(M_i)).$$

Unfortunately the least-square estimator is not robust against false matches, because the larger the residual $r_i$ is, the larger is its influence on the final estimate.

In order to reduce the influence of the feature outliers, i.e., features whose residuals are relatively large when the correct pose has been found, statisticians have suggested many different robust estimators. Among them, the two most popular are the *M-estimators* and the *least median-of-squares* (LMS) method, which have been used in many computer vision problems [Haralick 89] [Kumar 94] [Zhang 95].

The LMS technique, suggested by Rousseeuw and Leroy [Rousseeuw 87], consists of minimizing the median of the squared residuals:

$$\min_{\mathbf{p}} \operatorname{med}_i r_i^2, \tag{1}$$

where $\operatorname{med}_i x_i = x_{\lceil n/2 \rceil}$ if $x_1 \leq x_2 \leq x_n$.

Minimizing the median ignores the errors of the largest-ranked half of the data elements. Thus, this method is able to handle data sets which contain less than 50% outliers. However, as only a part of the data is considered, the LMS is not very accurate. Moreover, such an estimation is expensive, because LMS cannot be reduced to a straightforward formula.

We therefore prefer to use the *M-estimation* technique, developed by [Huber 81], which minimizes the sum of a function of the residuals:

$$\min_{\mathbf{p}} \sum_{i=1}^{n} \rho(r_i), \tag{2}$$

where $\rho$ is a continuous, symmetric function with minimum value at zero. Its derivative $\psi(x)$ is called the *influence function* because it occurs as a weighting function in optimization (2). These functions are very efficient, but are not suited to cases where the presence of outliers in the data is too great (experimentally, it must be kept below approximately 20%). Table 1 lists three commonly used $\rho$ functions and their derivatives. Among these estimators, some are more restrictive than others: when Tukey's influence function is null for residuals larger than a threshold $c$, Cauchy's remains larger than zero while decreasing, whereas Huber's remains constant, equal to $c$. We therefore prefer to use Tukey's function, which is restrictive enough to suppress the influence of outliers while still taking all the data into consideration (by contrast with LMS).

Minimization (2) can be performed by standard techniques using an initial estimate of $\mathbf{p}$: a very simple approach like Powell's method [Press 88] proved to be sufficient in our case, and relatively fast to compute (for temporal registration, the initial estimate of $\mathbf{p}$ is the pose computed for the previous frame).

| type | $\rho(x)$ | $\psi(x)$ | graph of $\psi(x)$ |
|---|---|---|---|
| Meansquares | $x^2/2$ | $x$ |  |
| Huber $\begin{cases} \text{if } |x| \le c \\ \text{if } |x| > c \end{cases}$ | $\begin{cases} x^2/2 \\ c(|x| - c/2) \end{cases}$ | $\begin{cases} x \\ c * \text{sgn}(x) \end{cases}$ |  |
| Cauchy | $\frac{c^2}{2} \log\left(1 + \left(\frac{x}{c}\right)^2\right)$ | $\dfrac{x}{1 + \left(\frac{x}{c}\right)^2}$ |  |
| Tukey $\begin{cases} \text{if } |x| \le c \\ \text{if } |x| > c \end{cases}$ | $\begin{cases} \frac{c^2}{6}\left[1 - \left(1 - \left(\frac{x}{c}\right)^2\right)^3\right] \\ c^2/6 \end{cases}$ | $\begin{cases} x\left(1 - \left(\frac{x}{c}\right)^2\right)^2 \\ 0 \end{cases}$ |  |

**Table 1.** A few commonly used M-estimators.

## 2.2   Computing Viewpoint from Curve Correspondences

This problem is much more difficult because point-to-point correspondences between the 3D curves and the 2D curves are not available. Let:

- $C_i$ be a 3D curve, described by the chain of 3D points $\{M_{i,j}\}_{1 \le j \le l_i}$ (note that $C_i$ can be any 3D feature, including points and lines);

- $c_i$ be the projection of $C_i$ in the image plane, described by the chain of 2D points $\{m_{i,j}\}_{1 \le j \le l_i}$, where $m_{i,j} = Proj(\mathbf{R}M_{i,j} + \mathbf{t})$ ($Proj$ denoting the projection of a 3D point in the image plane);

- $c_i'$ be the detected curve (tracked curve) corresponding to $C_i$, described by the chain of 2D points $\{m_{i,j}'\}_{1 \le j \le l_i'}$.

A simple solution would be to perform a one-stage minimization

$$\min_{\mathbf{P}} \sum_{i,j} \rho(d_{i,j}) \tag{3}$$

where $d_{i,j} = Dist(m_{i,j}', c_i)$ ($Dist$ being a function which approximates the Euclidean distance from a point to a contour) and $\rho$ is a positive, symmetric function, used to reduce the influence of erroneous parts.

Unfortunately, this method is unsatisfactory because it merges all the features into a set of points, and makes no distinction between local errors (when a feature is only partially well localized) and gross errors (when the position of a feature is completely erroneous). These two kinds of errors are not identical, and failing to treat them separately leads to a great loss of robustness and accuracy.

By contrast, we propose to perform a robust estimation in a two-stage process: a *local stage*, which computes a robust residual for each feature, and a *global stage* which minimizes a robust function of these residuals. The local stage reduces the influence of erroneous sections of the contours (features 1 and 4 in Figure 1c), whereas the global stage discards the *feature outliers, i.e.,* contours which are completely erroneous, or which contain too large a portion of erroneous points (feature 5 in Figure 1c).

**The Local Stage**

The aim of this stage is to reduce the influence of erroneous sections of the features. In order to perform this task, the residual error $r_i$ of curve $C_i$ is computed by a robust function of the distances $\{d_{i,j}\}_{1 \leq j \leq l'_i}$. We could take $r_i$ equal to the median of the $\{d_{i,j}\}$, but once again, this method can lead to a local minimum, in which only a part of a projected feature is superimposed on the corresponding 2D feature. That's why we prefer to use the M-estimation technique by taking

$$r_i^2 = \frac{1}{l'_i} \sum_{j=1}^{l'_i} \rho(d_{i,j}). \tag{4}$$

This estimate must not be too restrictive, for the reason just cited. We have hence chosen Huber's function for the local stage, which has proved to be a good choice in our experiments.

**The Global Stage**

Once a robust residual has been computed for each feature, the viewpoint is computed in a robust way by minimizing

$$\min_{\mathbf{p}} \sum_{i=1}^{n} \rho(r_i),$$

where $\rho$ is the Tukey function.

**Discarding Feature Outliers**

The detection of feature outliers can now be performed easily: since they should not have influenced the estimation, their residual must be much larger than those of the other features. We therefore only have to compare them with the standard deviation of all the residuals: if $r_i > 2.5 \hat{\sigma}$ (where $\hat{\sigma}$ is computed in a robust way [Kumar 94]), then the feature is discarded.

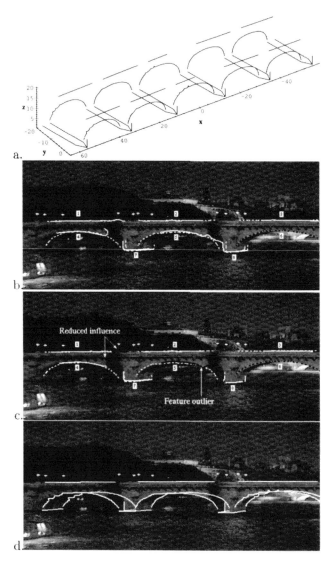

**Figure 1.** Temporal registration.
(a): Wireframe model of the object to be registered (the bridge). (b): Tracking of image features. White lines correspond to the tracked features, white dashed lines to the projection of their 3D correspondents in the previous frame, and black dashed lines to the features not (yet) used. (c): Robust pose computation. Sections for residuals greater than $c$ and for feature outliers are drawn in black. (d): Re-projection of the model.

In order to refine the pose, we can now perform a least-square estimation on the retained features ($r_i$ still being given by Equation 4).

The issue of maintaining registration over time (replacing feature outliers, integrating newly appearing features, etc.) is detailed in [Simon 98].

## 3   Improving the Viewpoint Computation

Computing the viewpoint from 2D/3D correspondences allows us to minimize the reprojection error for the model features. Unfortunately, if we want to add virtual objects in a scene area which does not contain (or contains few) model points, the reprojection error in this area is likely to be large. As an example, consider the scene depicted in Figure 2, in which we want to add a virtual car passing behind the statue (on the right). The viewpoint has been computed from 3D features belonging to the opera (the building in the back of the scene). The epipolar lines for the points drawn in Figure 2a are shown in Figure 2b: they pass far from the corresponding points. This proves that the viewpoint is not computed with sufficient accuracy. As a consequence, the projection of the virtual object in the scene is likely to be incorrect. Moreover, occluding objects which stand in front of the virtual ones will be erroneously reconstructed. Therefore, the occlusions between the virtual objects and the real scene cannot be solved properly, which reduces dramatically the realism of the image composition.

In order to improve the accuracy of the viewpoint, we use key-points that can be matched easily in two consecutive images. Since the relationship between two matched points is a function of the camera motion and of the intrinsic parameters, the viewpoint can be improved by minimizing a cost function which encompasses the reprojection error as well as the matching error between two frames. Since the key-points do not generally correspond to the model points, the viewpoint computation will be improved through these 2D correspondences.

**Figure 2.** Epipolar lines corresponding to the points drawn in (a); with the RPC algorithm (b); with the mixing method (c) with the mixed method.

Section 3.1 describes the way we extract key-points. Section 3.2 presents the cost function we use to improve the viewpoint. Significant results are shown in Section 4.

## 3.1  Extracting and Matching Key-points

Key-points (or interest points) are locations in the image where the signal changes two-dimensionally: corners, T-junctions, locations where the texture varies significantly, etc. Approaches for detecting key-points can be broadly divided in two groups: the first group involves first extracting edges and then searching for points that have maximum curvature; the second group works directly on the grey-level image. Since the edges are already used in the viewpoint computation, we resort to the second approach, which provides us with interesting texture points which are not yet used. We use the approach developed by Harris and Stephens [Harris 89]: they use the autocorrelation function of the image to compute a measure which indicates the presence of an interest point. More precisely, the eigenvalues of the matrix

$$\begin{bmatrix} I_x^2 & I_x I_y \\ I_x I_y & I_y^2 \end{bmatrix}$$

are the principal curvatures of the auto-correlation function. If these values are high, a key-point is declared.

We still have to match these key-points between two consecutive images. Numerous works use correlation techniques to achieve this task [Zhang 95]. These methods are well suited when the motion in the image is roughly a translation, but they are unable to cope with rotations and scale changes. That is the reason we prefer to use the matching approach developed in [Schmid 97]: each key-point is characterized locally by a vector of differential invariants under the group of displacements. For example, the vector of differential invariants up to second order is

$$\begin{bmatrix} I \\ I_x^2 + I_y^2 \\ I_{xx} I_x^2 + 2 I_{xy} I_x I_y + I_{yy} I_y^2 \\ I_{xx} + I_{yy} \\ I_{xx}^2 + 2 I_{xy} I_{yx} + I_{yy}^2 \end{bmatrix}.$$

The invariance of the vector makes the matching stage easier even in case of important geometric transformations. Moderated scale changes can also be considered. The interested reader can find further details on the computation of the local invariants in [Schmid 97]. The key-points are then matched according to a measure of similarity between the invariant vectors. Neighboring constraints are also used to make the matching easier (close key-points must have a similar disparity).

## 3.2 Mixing 3D Knowledge and Point Correspondences

Given the viewpoint $[\mathbf{R}_k, \mathbf{t}_k]$ computed in a given frame $k$, we now explain how we compute the viewpoint in the next frame, $k + 1$, using the 3D model as well as the matched keypoints $(q_1^i, q_2^i)_{1 \leq i \leq m}$. Before describing the cost function to be minimized, we recall in some detail the relationships between two matched key-points $m_1, m_2$ and the two viewpoints $[\mathbf{R}_k, \mathbf{t}_k]$ and $[\mathbf{R}_{k+1}, \mathbf{t}_{k+1}]$. Let $\Delta\mathbf{R}, \Delta\mathbf{t}$ be the relative displacement of the camera between frames $k$ and $k + 1$. Let $A$ be the intrinsic matrix of the camera:

$$A = \begin{bmatrix} k_u f & 0 & u_0 \\ 0 & k_v f & v_0 \\ 0 & 0 & 1 \end{bmatrix}.$$

Let $q_1$ and $q_2$ be the images of a 3D point $M$ on the cameras. Their homogeneous coordinates are denoted $\tilde{q}_1$ and $\tilde{q}_2$. We then have the fundamental Equation [Luong 93]

$$\tilde{q_2}^t A^{-1}{}^t \Delta\mathbf{T}\Delta\mathbf{R} A^{-1} \tilde{q_1} = 0,$$

where $\Delta\mathbf{T}$ is an antisymmetric matrix such that $\Delta\mathbf{T}x = \Delta\mathbf{t} \wedge x$ for all $x$. $F = A^{-1}{}^t \Delta\mathbf{T}\Delta\mathbf{R} A^{-1}$ is called the fundamental matrix.

Then, a simple way to improve the viewpoint computation using the interest points is to minimize

$$min_{\mathbf{R}_{k+1}, \mathbf{t}_{k+1}} \left( \frac{1}{n} \sum_{i=1}^{n} \rho_1(r_i) + \frac{\lambda}{m} \sum_{i=1}^{m} \rho_2(vi) \right), \tag{5}$$

where

- $r_i$ is the distance in frame $k + 1$ between the tracked features and the projection of the model features;

- $v_i = \sqrt{\frac{1}{(F\tilde{q_1^i})_1^2 + (F\tilde{q_1^i})_2^2} + \frac{1}{(F^t\tilde{q_2^i})_1^2 + (F^t\tilde{q_2^i})_2^2}} |\tilde{q_2^i}^t F\tilde{q_1^i}|$ measures the quality of the matching between $q_i$ and $q_i'$ [Luong 93];

- $\rho_1$ and $\rho_2$ are M-estimators. Note that the use of an M-estimator for the key-points correspondences is not essential: since the key-points are significant points in the image, false matches are unusual.

The $\lambda$ parameter controls the compromise between the closeness to the available 3D data and the quality of the 2D correspondences between the key-points. We use $\lambda = 2$ in our practical experiments.

The minimum of Equation 5 is computed by using an iterative algorithm for minimization such as Powell's algorithm.

## 4   Results

Figure 3 shows the result of using the mixing algorithm to incorporate a virtual car in a video sequence. A video of Stanislas Square in the city of Nancy, France, has been shot from a car driving around the square. Our aim is to incorporate a virtual car passing on the square. It is worth noting here that the 3D data available for the scene only concern the opera. No 3D model is available for the statue. The virtual car is incorporated in the foreground of the scene, on the ground near the statue.

Figure 3a shows the projection of the opera model in the image accomplished by using the RPC algorithm alone with the car incorporation. The projection error seems minimal but the car seems to hover! This result is not really surprising since the viewpoint computation has been performed with nearly coplanar points in the background of the scene. The key-points extracted from the scene and the matching arrows are shown in Figure 3b. The result of the mixing algorithm is shown in Figure 3c. The car and the scene are combined seamlessly, and the realism of the composition is very good.

Figure 4 exhibits the viewpoint evolution for the RPC algorithm and the mixing method. In the considered sequence, the camera is passing on the opposite side of the opera. The motion of the camera is then a translation along the $x$-axis. Figure 4 proves the efficiency of the mixing method: the $t_y$ and $t_z$ coordinates are nearly constant, whereas the $t_x$, $\alpha$, $\beta$, and $\gamma$ parameters evolve slowly.

Other significant results with video image sequences can be seen at URL http://www.loria.fr/~gsimon/videos.html.

## 5   Conclusion

We have presented a robust and accurate registration method which allows us to combine the real and the virtual worlds seamlessly. One of the main advantages of our approach is that it performs pose computation over the sequence in a completely autonomous manner. The accuracy of the pose computation derives from the combined use of 3D-2D correspondences in

**Figure 3.** Result of the mixing algorithm: (a) the image composition using the viewpoint computed with the RPC algorithm (b) the extracted keypoints (c) image composition with the result given by the mixing algorithm.

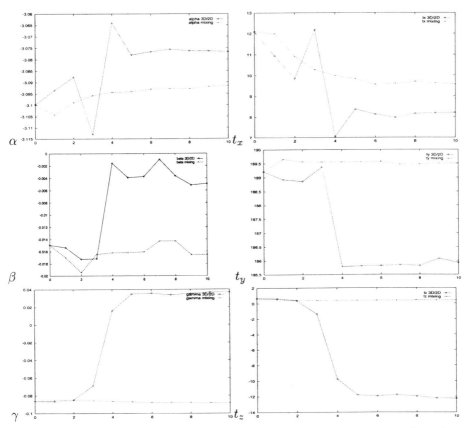

**Figure 4.** Evolution of the motion parameters (Euler angles and translation) for the RPC and the mixing algorithm.

an image and 2D-2D correspondences in two consecutive frames: indeed, the use of 2D-2D correspondences allows us to bring some kind of spatial information to the scene in areas where 3D-model features are missing. As a result, our method only requires a limited number of 3D features in order to be effective.

Currently, the time needed to process one frame is about 25 s (for five 3D curves and 100 key-points on an UltraSparc 143-MHz). However, we can greatly improve the speed of our algorithm by processing both the pose computation and the key-points extraction in a parallel way. In this way, we think that our algorithm is amenable to real time.

Future works will concern the automatic determination of the intrinsic camera parameters. Indeed, these parameters are currently computed off-line before shooting the sequence. Since the zoom of the camera may change during shooting, it would be interesting to compute dynamically the intrinsic camera parameters.

# References

[Berger 94]     M.-O. Berger, "How to Track Efficiently Piecewise Curved Contours with a View to Reconstructing 3D Objects," *Proceedings of the 12th International Conference on Pattern Recognition*, Jerusalem, Israel, Vol. 1, pp. 32–36, 1994.

[Berger 96]     M.-O. Berger, C. Chevrier, and G. Simon, "Compositing Computer and Video Image Sequences: Robust Algorithms for the Reconstruction of the Camera Parameters," *Computer Graphics Forum, Conference Issue Eurographics '96*, Poitiers, France, Vol. 15, pp. 23–32, August 1996.

[Berger 97]     M.-O. Berger, "Resolving Occlusion in Augmented Reality: a Contour-based Approach without 3D Reconstruction," *Proceedings of IEEE Conference on Computer Vision and Pattern Recognition*, Puerto Rico, USA, pp. 91–96, June 1997.

[Dementhon 95] D. Dementhon and L. Davis, "Model Based Object Pose in 25 Lines of Code," *International Journal of Computer Vision*, Vol. 15, pp. 123–141, 1995.

[Ertl 91]       G. Ertl, H. Müller-Seelich, and B. Tabatabai, "MOVE-X: A System for Combining Video Films and Computer Animation," *Eurographics*, pp. 305–313, 1991.

[Faugeras 86]   O. D. Faugeras and G. Toscani, "The Calibration Problem for Stereo," *Proceedings of IEEE Conference on Computer Vision and Pattern Recognition,* Miami, FL, pp. 15–20, 1986.

[Faugeras 86]   O. Faugeras, *Three-Dimensional Computer Vision: A Geometric Viewpoint*, Artificial Intelligence, MIT Press, Cambridge, MA, 1993.

[Feldmar 97]    J. Feldmar, N. Ayache, and F. Betting, "3D-2D Projective Registration of Free Form Curves and Surfaces," *Computer Vision and Image Understanding*, Vol. 65, No. 3, pp. 403–424, 1997.

[Ferri 93]      M. Ferri, F. Mangili, and G. Viano, "Projective Pose Estimation of Linear and Quadratic Primitives in Monocular Computer Vision," *CVGIP: Image Understanding*, Vol. 58, No. 1, pp. 66–84, July 1993.

[Haralick 89]     R.M. Haralick, H. Joo, C.N. Lee, X. Zhuang, V.G. Vaidya, and
                  M.B. Kim, "Pose Estimation from Corresponding Point Data,"
                  *IEEE Transactions on Systems, Man, and Cybernetics*, Vol. 19,
                  No. 6, pp. 1426–1446, 1989.

[Harris 89]       C. Harris and M. Stephens, "A Combined Corner and Edge De-
                  tector," *Proceedings of 4th Alvey Conference*, Cambridge, UK,
                  pp. 147–151, August 1988.

[Huber 81]        P.J. Huber, *Robust Statistics*, Wiley, New York, 1981.

[Kriegman 90]     D. Kriegman and J. Ponce, "On Recognizing and Positioning
                  Curved 3D Objects from Image Contours," *IEEE Transactions
                  on PAMI*, Vol. 12, No. 12, pp. 1127–1137, December 1990.

[Kumar 94]        R. Kumar and A. Hanson, "Robust Methods for Estimating Pose
                  and a Sensitivity Analysis," *CVGIP: Image Understanding*, Vol.
                  60, No. 3, pp. 313–342, 1994.

[Luong 93]        Q.-T. Luong, R. Deriche, O. Faugeras, and T. Papadopoulo,
                  "On Determining the Fundamental Matrix: Analysis of Different
                  Methods and Experimental Results," *Rapport de recherche* 1894,
                  INRIA, 1993.

[Press 88]        W.H. Press, B.P. Flannery, S.A. Teukolsky, and W.T. Vetter-
                  ling, *Numerical Recipes in C, The Art of Scientific Computing*,
                  Cambridge University Press, Cambridge, UK, 1988.

[Ravela 96]       S. Ravela, B. Draper, J. Lim, and R. Weiss, "Tracking Object
                  Motion Across Aspect Changes for Augmented Reality," *ARPA
                  Image Undertanding Worshop*. Palm Springs, August 1996.

[Rousseeuw 87]    P. Rousseeuw and A. Leroy, *Robust Regression and Outlier De-
                  tection*, Wiley Series in Probability and Mathematical Statistics,
                  Wiley, New York, 1987.

[Schmid 97]       C. Schmid and R. Mohr, "Local Grayvalue Invariants for Image
                  Retrieval," *IEEE Transactions on PAMI*, Vol. 19, No. 5, pp.
                  530–535, August 1997.

[Simon 98]        G. Simon and M.-O. Berger, "A Two-stage Robust Statistical
                  Method for Temporal Registration from Features of Various
                  Type," *Proceedings of 6th International Conference on Com-
                  puter Vision*, Bombay, India, pp. 261–266, January 1998.

[State 96]        A. State, G. Hirota, D. Chen, W. Garett, and M. Livingston,
                  "Superior Augmented Reality Registration by Integrating Land-
                  mark Tracking and Magnetic Tracking," *Computer Graphics
                  (Proceedings Siggraph)*, New Orleans, pp. 429–438, 1996.

[Tomasi 92]   C. Tomasi and T. Kanade, "Shape and Motion from Image Streams under Orthography: A Factorization Method," *International Journal of Computer Vision*, Vol. 9, No. 2, pp. 137–154, 1992.

[Uenohara 96]   M. Uenohara and T. Kanade, "Vision Based Object Registration for Real Time Image Overlay," *Journal of Computers in Biology and Medicine*, 1996.

[Zhang 95]   Z. Zhang, R. Deriche, O. Faugeras, and Q. Luong, "A Robust Technique for Matching Two Uncalibrated Images Through the Recovery of the Unknown Epipolar Geometry," *Artificial Intelligence*, Vol. 78, pp. 87–119, October 1995.

# A Fast and Robust Line-based Optical Tracker for Augmented Reality Applications

Didier Stricker, Gudrun Klinker, and Dirk Reiners

**Abstract.** *The correct alignment of real and virtual objects is one of the key technological problems in augmented reality (AR). The user's point of view must be determined accurately and tracked over time.*

*In this paper, we present an optical tracking system, focusing on real-time performance and robustness, including rapid recovery from unpredictable abrupt motions, as well as stable handling of partial occlusions of tracking targets. The algorithm tracks linear features of landmarks and other objects. For increased performance, we do not extract entire lines but rather use only a few sample points. The 3D motion recovery is done iteratively using robust estimation methods.*

*The system is able to run on standard hardware with a high frame rate (20-25 Hz on an $O_2$) and is robust enough to track a hand-held or head-worn camera.*

*We finish with a presentation of AR applications in the domains of exterior construction, 3D user interfaces, and games.*

## 1  Introduction

Augmented reality (AR) integrates computer-generated information into views of the real world, and is potentially useful in many application domains. It can assist users in maintenance and repair tasks for complex devices [Rose 95]. In exterior construction, the CAD model can be overlaid on video of the construction site and thus help during the planning and the construction phases [Klinker 98]. Furthermore, information from other sensors can be directly merged into the user's view in order to provide supplementary information on-line in various operations, as in medical and surgical contexts [Bajura 92] [Grimson 95].

129

## 1.1    Related Work

The registration of the virtual objects with the real world is a crucial problem in AR. In order to preserve the coherence of the augmented scene, the different sensors have to be carefully calibrated [Tuceryan 95]. Commercial tracking devices, such as magnetic trackers [Rose 95] and active LED-systems [Starner 97] [Webster 96] may be used, but since the precision of these devices and the extent of their working space are not sufficient for many AR applications, many researchers are now developing computer-vision-based methods. Typically, special landmarks are placed in the scene and are used to calibrate and track the camera [Hoff 96], [Bajura 95], [Cho 97], and [Mellor 95]. Other approaches use image features directly, but, require the initial pose to be given interactively [Ravela 95]; in some approaches, the system automatically finds correspondences in already calibrated images, and derives the camera pose from them [Uenohara 95]. In real-time applications, the measurement frequency plays a crucial role, thereby limiting the system complexity and the set of motions that can be tracked [Azuma 95]. Improvements in robustness can be achieved by fusing the image measurements with the information from such other sensors as inertial trackers [Azuma 94] or magnetic trackers [State 96].

In computer vision, 3D camera motion is a classic topic often related to robotic and navigation tasks. A particularly successful approach physically models the motion and exploits the image measurements using a Kalman filter [Bar-Shalom 88], [Zhang 92], [Azarbayejani 95], and [Koller 97].

In our applications, we assume that the motion of head-worn cameras can be very abrupt and unpredictable because users can turn their heads very quickly, resulting in major shifts in the image. Thus, we do not use a motion model and favor the dynamic properties of the tracker. Our system has been influenced by the approaches related in [Kumar 94], [Weng 92b], and [Lowe 92], but we pay more attention to the dynamic and real-time aspects by optimizing the feature search process and developing a new motion recovery algorithm.

## 1.2    Paper Outline

The goal of our system is to track a hand-held or head-worn camera with a high frame rate using standard hardware. We present an algorithm showing the feasibility of performing highly precise and fast 3D tracking using computer vision methods. After an overview of our system, we describe the details of the initialization and the 3D tracking algorithms. Finally, we evaluate the system performance and present some applications.

# 2  System Overview

## 2.1  Requirements

The user should feel free in his movements and actions. This goal necessitates the following system features:

- *Automatic and fast initialization.* The system should not require users to interactively support the tracker initialization phase.

- *Real-time tracking.* The frame rate should be higher than 10 Hz, and ideally higher than 20 Hz. The goal is to minimize the dynamic error, i.e., the lag between the view of the real world and the rendering of the augmented view for the user. This condition is not only relevant for see-through systems (Figure 1) [Azuma 94] but also for video feed-through applications or when the image augmentation is accomplished by another process. Fast tracking also minimizes the inter-frame difference and thus the searching distance between tracking features.

- *Error supervision and fast re-initialization.* The system should guarantee minimal error margins for the alignment of the augmentation with the real world. When it fails, it should be able to reinitialize itself with minimal disturbance for the user, i.e., without user interaction and in a very short time.

To achieve the above requirements, we use landmarks for fully automatic calibration, and we optimize the tracking system by developing appropriately robust algorithms.

**Figure 1.** Camera used as a tracker device

## 2.2    System Components

Our approach is model-based, using dark rectangles on a bright background. In order to be identifiable independently of the current field of view, the rectangles contain one or two rows of red marks defining a four-bit (or eight-bit) code. This kind of bar code gives the system significant flexibility, due to the full identification of each landmark and the potentially high number of such IDs (up to 255).

The initialization is achieved in two steps: First, the corners of the landmarks are extracted and a rough calibration is computed. Next, after back-projecting the landmarks model into the image, the corner positions are refined, enabling an accurate re-calibration. The subsequent image-to-image feature tracking is composed of a position prediction and a target re-detection phase, followed by the update of the 3D position and the corresponding error estimation (see Figure 2).

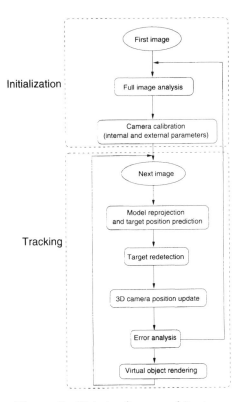

**Figure 2.** Global software architecture

# 3 Initialization and Calibration

## 3.1 Robust Rectangle Detection

The rectangles are first extracted as blob candidates, which are defined by three points assumed to lie on three of the four edges. The points are found by scanning the image every $n$ lines. Two consecutive image gradients, white-to-dark and dark-to-white, define the left and right borders of the blob. Starting from the center between them, the third point is found by looking vertically for a strong image gradient. We then check the homogeneity of the blob and its average level of blackness along the scanned segments of the blob.

We then follow the blob contour. Starting from a border pixel, we calculate the gradient norm of all eight of its neighbors. To choose the next pixel, we use a criterion $C_{ij}$ defined below. It combines the neighbor gradient, the actual gradient, and the displacement vector.

$$C_{ij} = (\vec{\nabla} I(P_{00}).\vec{v_{ij}}) \, \| \vec{\nabla} I(P_{ij}) \|$$

where $\vec{v_{ij}}$ is the unit vector orthogonal to the displacement from the current pixel at position $(0,0)$ to the neighbor pixel at position $(i,j)$; $I(P_{00})$, and $I(P_{ij})$ represent the the intensity value of the pixel $P$ at position $(0,0)$ and $(i,j)$; $(i = -1, 1; j = -1, 1)$

To fit rectangles to each blob, the contour samples are classified into four clusters according to their gradient direction, using a standard ISO-data algorithm. After a statistical homogeneity test, i.e., a comparison of cluster sizes, mean values, and standard deviations of the gradient norm

**Figure 3.** Landmark detection and identification

and direction, we fit a straight line to the edges. The intersection of neighboring lines determines the corner points of the rectangle.

## 3.2   Identification

To relate each blob uniquely to one of the squares described in a 3D model
of the environment, we examine the labeling area within each blob, interpreting the line of red markers as a binary code. The code is read by
sampling the line of marks and correlating it with templates representing
the encoding of all possible identification numbers. We use the zero mean
normalized correlation and consider the rectangle to be identified if the
highest score can be defined without ambiguity. This method has proven
to be very robust and also to work well with low-quality cameras (e.g., an
Indy-Cam) and under poor illumination conditions.

## 3.3   Calibration

A priori we do not know what kind of virtual object will be inserted in the
scene; for this reason, we choose the most general camera model, the pin-
hole model. The pose of the camera is defined by the rigid transformation:

$$\begin{pmatrix} x_c \\ y_c \\ z_c \end{pmatrix} = R \begin{pmatrix} x_w \\ y_w \\ z_w \end{pmatrix} + T$$

And the relation to the image coordinate system is:

$$u = f.s_x \frac{x_c}{z_c} + c_{x0}$$

$$v = f.s_y \frac{y_c}{z_c} + c_{y0}$$

We calculate by full calibration the internal ($f$, $s_x$, $s_y$, $c_{x0}$, $c_{x0}$) and
external parameters ($R$, $T$) of the camera using the algorithms described
in [Weng 92a] [Tsai 86]. The lens distortion is actually not compensated.

## 4   Tracking Algorithm

The main steps of the tracking algorithm are the target position prediction,
the re-detection, and finally the motion recovery. Each point is detailed in
this section.

## 4.1   Prediction

Due to the random character of user head motion, camera motions can be very erratic. We thus limit motion prediction to very basic, linear 2D extrapolations in the image. Our experiments have shown that such fast linear approximations by far outperform in robustness complex, physically more correct, 3D motion models. 3D models are much more time-consuming to compute, and thus systems that use them have to deal with much longer intervals of measurement.

The velocities of the corners of each square are calculated individually from their current and previous positions and are used to predict their approximate location in the next image. This specifies the search areas within which their exact location will be determined.

## 4.2   Target Re-detection

In the tracking process, linear features of the targets need to be found in the new image. To avoid time-consuming feature extraction processing across the entire image, we work locally along search segments. The image, Figure 4(a), shows those segments, which are defined by lines that are perpendicular to the re-projected model square's sides. In practice, we limit their directions to the image row, column, and diagonal.

We then compute the gradient along each one using a Gaussian kernel, and localize the maximum with sub-pixel accuracy by interpolating to the second order. These new points determine implicitly the new position of the landmarks in the image.

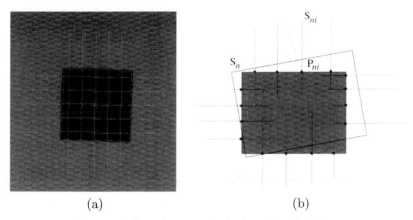

(a)                                     (b)

**Figure 4.** Search segments for target re-detection

Let $S_n$ denote the $n^{th}$ segment of the re-projected landmarks model in the image (Figure 4(b)), and let $S_{ni}$ denote the $i^{th}$ search segment of $S_n$. The point with maximum gradient found along $S_{ni}$ is labeled $P_{ni}$.

## 4.3   Discussion

Pose and motion estimation is usually done with well-defined primitives like points, lines, and conics [Zhang 92] [Kumar 94].

From the precedent processing, a line could be fitted, through the detected points $P_{ni}$ of a same segment $S_n$ in order to determine the landmark side. The intersection of them would define the square corner in the image, and we could apply points algorithm like those presented in [Klinker 98].

However, we have found formal line-fitting and corner-computation approaches not to be very robust in the presence of partial occlusions, fast motion, and noise: since the line must be extrapolated from a small number of points, and since some points $P_{ni}$ are discarded or poorly localized (because of occlusions or fast motions), the derived location of lines is often imprecise.

On the other hand, most of the originally detected points $P_{ni}$ are well located on the target border. The idea of the following algorithm is to exploit directly that information rather than first to derive new (uncertain) primitives from them. In this way, our method is statistically more robust, because we work with a larger number of independent primitives.

## 4.4   Recovery of the motion parameters

To recover the motion, we have to define a constraint between the newly detected points $P_{ni}$ and the projected model in the image.

The position of the camera should be updated, so that the re-projected segment $S_n$ lies on the landmarks border in the image. That means also that the distance from the points $P_{ni}$ to this segment is null.

We express the distance constraint formally as follows: If $U_n$ is an orthogonal unit vector of the segment $S_n$, and M is an arbitrary point of $S_n$, we have the equality:

$$U_n.(P_{ni} - M) = 0$$

Thus, we recover iteratively the position of the camera by taking the previously determined parameters (three rotation angles and the three translation components) as initial values and minimizing the objective function $F$:

$$F = \sum_{n=1}^{N} \sum_{i=1}^{S} (U_n.(P_{ni} - M))^2$$

with $N$ being the number of segments in the model and $S$ being the number of samples per segment $S_n$.

This criterion minimizes the orthogonal distance of the points $P_{ni}$ to the corresponding model segment $S_n$.

## 4.5   Minimal Number of Samples

Camera pose is defined by six parameters [Haralick 94]. Thus, pose can in principle be recovered from three points, with each providing two constraints, one per coordinate axis. Yet the points $P_{ni}$ are defined along the segments $S_{ni}$. They are constrained only in the $S_{ni}$ direction. Each point $P_{ni}$ therefore provides only one constraint, so that the algorithm requires at least six points, not all collinear.

## 5   Robustness

We have explored several ways of improving the tracker to make it robust and flexible enough for real and difficult applications. The critical issues are dealing robustly with occlusion, integration of already existing scene lines, and extension to hybrid tracking.

## 5.1   Partial Occlusions and Outliers

If marks are partially occluded, some of the points $P_{ni}$ are not determined correctly. To limit their influence on the pose estimation process, we use robust statistic methods, such as the M-estimator [Meer 91], [Press 92], and [Zhang 97].

In principle, we can also use outlier detection techniques. Yet, practical experience has shown that outlier detection is not advantageous at this stage: Since the precision of the camera calibration is inherently related to the location of the currently visible squares in the scene, new squares that become visible during head rotations may not fit the current camera model well. Declaring such misfits to be outliers means that they are excluded from influencing the camera model to adapt to scene information beyond the initial field of view.

**Figure 5.** Camera tracking without landmarks

## 5.2   Integration of Scene Lines

The presented algorithm is not only valuable for the designed targets but can be extended to arbitrary polyhedral objects.

The advantage is twofold: First the user can go away from the targets after the initialization phase (see Figure 5), and the camera will still track in 3D. Secondly it stabilizes the tracking when only a few targets are present in the image.

## 5.3   Externally Available Information and Hybrid Tracking

For monitor-based AR, the camera is often placed on a tripod near the user (see Figure 6). Camera motion is then typically limited to rotations around the fixed point on the tripod. In some cases, other sensors, such as GPS (Global Positioning System) for wide environment, can determine the camera position.

**Figure 6.** Monitor-based AR: the camera translation is constant

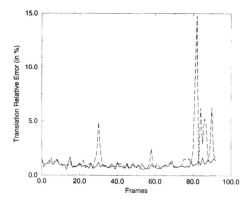

**Figure 7.** Relative error measurement

In such scenarios, we estimate only the rotation parameters; the translation is considered constant or is provided by the other sensors. The result is an appreciable reduction of computation time combined with increased robustness and stability. In Section 7, we demonstrate the use of such external knowledge for an application in an exterior construction site.

# 6   System performance

Figure 7 shows a comparison between the new line-based algorithm and a previous corner-based approach (dashed line) (see Section 4.3 as well as [Klinker 98]). Using a synthetic image sequence of 95 pictures, we calculate the relative errors in the camera position for the two algorithms.

The spikes indicate situations when some targets were only partially in the field of view. They are not taken into account by the preceeding algorithm. The position estimation error grows so that a new initialization becomes necessary. The proposed algorithm maintains the relative error under 2.5%.

| Machine | Initialization | Tracking |
|---------|----------------|----------|
| $O_2$ | 5 Hz | 23 Hz |
| Indy 5000 | 3 Hz | 12 Hz |

**Table 1.** Real time performance

The real-time performance displayed in the above table was measured with four landmarks in an image video of size $768 \times 576$ pixels.

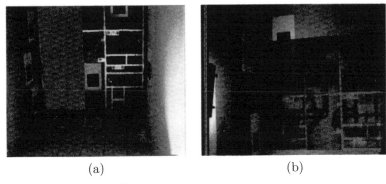

(a)                                                    (b)

**Figure 8.** Construction planning and maintenance

# 7 Applications

The following applications illustrate the potential of AR and the robustness of the system we have presented.

## 7.1 Exterior Construction

In Figure 8(a), taken during the construction phase of a building, a virtual grid and a wall are added into the scene to show the next planned step. The original picture of Figure 8(b) was taken after construction of the wall. The user sees the water pipes "through" the wall using "X-ray augmentation," and can therefore see their real positions during a maintenance task.

In the next scenario (Figure 9), the position (translation) of the camera was determined by a differential GPS, and the 3D position of the squares was measured with a laser pointer. The system is then able to recover correctly and reliably the camera rotation parameters during a panorama swing, augmenting images with a virtual wall.

**Figure 9.** Hybrid tracking with GPS

(a)                                    (b)

**Figure 10.** Physical manipulation and interaction

## 7.2 Model Presentation and Physical Interaction

Virtual models can be presented in 3D in the real environment using stereo rendering on a see-through, head-mounted display (Figure 10(a)).

The virtual object and the real object (the cardboard) are components of one entity, a *"mixed object."* The manipulation is then very intuitive, because the user interacts with a virtual object using physical means.

## 7.3 Game

The Tic-Tac-Toe game in Figure 10(b) explores different interaction schemes in an augmented environment. The user sets his stone and pushes a virtual "GO" button. The computer analyzes the scene, localizes the stone (by color segmentation), places its next virtual cross, and, on a virtual panel, instructs the user to continue. All interactions occur directly in the real world – away from the keyboard.

If the camera is static, a background subtraction enables scene change detection—for instance, the moving hands of the Tic-Tac-Toe player in front of the camera. By initializing the $Z$-buffer, we can resolve the occlusion problem under the assumption that the moving real object are in foreground of the scene. In our experience, the dynamic occlusion handling greatly helps users to understand and interact intuitively with the augmented scene. This application runs at about 8 Hz.

## 8 Summary and Conclusions

In this article, we have presented a robust tracking algorithm developed for the efficient recovery of moderately fast camera motion. The system allows for experiments in AR with a hand-held or head-mounted camera. It does not require interactive support from the user and works on standard hardware.

Nevertheless, the camera motions are currently limited to registered areas with known landmarks, or predefined polyhedral objects. Further developments will integrate automatic detection of new image features [Shi 94], which can be reconstructed in 3D and used for tracking.

## Acknowledgments

Laboratory space and equipment were provided by the European Computer-industry Research Center (ECRC). Our research was partially funded by the European CICC and CUMULI projects.

## References

[Azarbayejani 95]  A. Azarbayejani and A.P. Pentland. "Recursive Estimation of Motion, Structure, and Focal Length," *PAMI*, Vol. 17, No. 6, pp. 562–575, June 1995.

[Azuma 94]  R. Azuma and G. Bishop. "Improving Static and Dynamic Registration in an Optical See-Through HMD," *Proceedings of SIGGRAPH '94*, Orlando, FL, pp. 197–204, July 1994.

[Azuma 95]  R. Azuma and G. Bishop. "A Frequency-Domain Analysis of Head-Motion Prediction," *Proceedings of SIGGRAPH '95*, Los Angeles, CA, pp. 401–408, August 1995.

[Bajura 92]  M. Bajura, H. Fuchs, and R. Ohbuchi. "Merging Virtual Objects with the Real World: Seeing Ultrasound Imagery with the Patient," *Computer Graphics*, Vol. 26, No. 2, pp. 203–210, July 1992.

[Bajura 95]  M. Bajura and U. Neumann. "Dynamic Registration Correction in Video-based Augmented Reality Systems," *IEEE Computer Graphics and Applications*, Vol. 15, No. 5, pp. 52–60, 1995.

[Bar-Shalom 88]  Y. Bar-Shalom and T.E. Fortmann. "Tracking and Data Association," Academic Press, 1988.

[Cho 97]  Y. Cho, J. Park, and U. Neumann. "Fast Color Fiducial Detection and Dynamic Workspace Extension in Video See-through Self-Tracking Augmented Reality," *Proceedings of the Fifth Pacific Conference on Computer Graphics and Applications*, Seoul, Korea, pp. 168–177, 1997.

[Grimson 95]    W.E.L. Grimson, G.J. Ettinger, S.J White, P.L. Gleason, T. Lozano-Perez, W.M Wells III, and R. Kikinis. "Evaluating and Validating an Automated Registration System for Enhanced Reality Visualization in Surgery," *Proceedings of CVRMed '95*, Nice, France, pp. 3–12, April 1995.

[Haralick 94]    R.M. Haralick, C.N. Lee, K. Ottenberg, and M. Nolle. "Review and Analysis of Solutions of the 3-Point Perspective Pose Estimation Problem," *IJCV*, Vol. 13, No. 3, pp. 331–356, December 1994.

[Hoff 96]    W.A. Hoff, K. Nguyen, and T. Lyon. "Computer Vision-based Registration Techniques for Augmented Reality," *Proceedings of Intelligent Robots and Computer Vision XV, SPIE*, Vol. 2904, Boston, MA, November 1996.

[Klinker 98]    G. Klinker, D. Stricker, and D. Reiners. "Augmented Reality for Exterior Construction Applications," W. Barfield and T. Caudell, eds., *Augmented Reality and Wearable Computers*, Lawrence Erlbaum Press, Hillsdale, NJ, to appear 1999.

[Koller 97]    D. Koller, G. Klinker, E. Rose, D. Breen, R. Whitaker, and M. Tuceryan. "Automated Camera Calibration and 3D Egomotion Estimation for Augmented Reality Applications," *Proceedings of CAIP '97*, Kiel, Germany, pp. 199–209, September 1997.

[Kumar 94]    R. Kumar and A.R. Hanson. "Robust Methods for Estimating Pose and a Sensitivity Analysis," *CVGIP-IU*, Vol. 60, No. 3, pp. 313–342, November 1994, ftp://cicero.cs.umass.edu /Text/kumar/cvgip-iu94.ps.Z

[Lowe 92]    D.G. Lowe. "Robust Model-Based Motion Tracking Through the Integration of Search and Estimation," *IJCV*, Vol. 8, No. 2, pp. 113–122, August, 1992.

[Meer 91]    P. Meer, D. Mintz, D.Y. Kim, and A. Rosenfeld. "Robust Regression Methods for Computer Vision: A Review," *International Journal of Computer Vision*, Vol. 6, No. 1, pp. 59–70, April 1991.

[Mellor 95]    J.P. Mellor. "Realtime Camera Calibration for Enhanced Reality Visualization," *Proceedings of Computer Vision, Virtual Reality and Robotics in Medicine (CVRMed '95)*, Nice, France, pp. 471–475, April 1995.

[Press 92]    W.H. Press, S.A. Teukolsky, W.T. Vetterling, and B.P. Flannery. *Numerical Recipes in C: The Art of Scientific Computing*, 2nd edition, Cambridge University Press, Cambridge, UK, 1992.

[Ravela 95]    S. Ravela, B. Draper, J. Lim, and R. Weiss. "Adaptive Track-
               ing and Model Registration Across Distinct Aspects," *Inter-
               national Conference on Intelligent Robots and Systems*, IEEE,
               Pittsburgh, PA, August 1995.

[Rose 95]      E. Rose, D. Breen, K.H. Ahlers, C. Crampton, M. Tuceryan,
               R. Whitaker, and D. Greer. "Annotating Real-World Objects
               Using Augmented Reality," *Proceedings of Computer Graph-
               ics: Developments in Virtual Environments*, Academic Press
               Ltd, London, pp. 357–370, January 1995.

[Shi 94]       J. Shi and C. Tomasi. "Good Features to Track," *Proceedings
               of the Conference on Computer Vision and Pattern Recogni-
               tion*, Los Alamitos, CA, IEEE Computer Society Press, pp.
               593–600, June 1994.

[Starner 97]   T. Starner, S. Mann, B. Rhodes, J. Levine, J. Healey, D.
               Kirsch, R.W. Picard, and A. Pentland. "Augmented Real-
               ity Through Wearable Computing," *Presence, Special Issue on
               Augmented Reality*, Vol. 6, No. 4, pp. 386–398, August 1997.

[State 96]     A. State, G. Hirota, D.T. Chen, B. Garrett, and M. Liv-
               ingston. "Superior Augmented Reality Registration by Inte-
               grating Landmark Tracking and Magnetic Tracking," *Proceed-
               ings of SIGGRAPH '96*, New Orleans, pp. 429–438, August
               1996.

[Tsai 86]      R.Y. Tsai. "An Efficient and Accurate Camera Calibration
               Technique for 3D Machine Vision," *Proceedings of CVPR*,
               IEEE, Miami, FL, pp. 364–374, 1986.

[Tuceryan 95]  M. Tuceryan, D. Greer, R. Whitaker, D. Breen, C. Cramp-
               ton, E. Rose, and K. Ahlers. "Calibration Requirements and
               Procedures for a Monitor-Based Augmented Reality System,"
               *IEEE Transactions on Visualization and Computer Graphics*,
               Vol. 1, No. 3, pp. 255–273, September 1995.

[Uenohara 95]  M. Uenohara and T. Kanade. "Vision-Based Object Registra-
               tion for Real-Time Image Overlay," *Proceedings of CVRMed
               '95*, Nice, France, pp. 13–22, April 1995.

[Webster 96]   A. Webster, S. Feiner, B. MacIntyre, W. Massie, and T.
               Krueger. "Augmented Reality in Architectural Construction,
               Inspection, and Renovation," *Proceedings of ASCE Third
               Congress on Computing in Civil Engineering*, Anaheim, CA,
               pp. 913–919, June 1996.

[Weng 92a]     J. Weng, P. Cohen, and M. Herniou. "Camera Calibration with
               Distortion Models and Accuracy Evaluation," *PAMI*. Vol. 14,
               No. 10, pp. 965–980, 1992.

[Weng 92b]     J. Weng, T.S. Huang, and N. Ahuja. "Motion And Struc-
               ture From Image Sequences," Vol. 93, Springer-Verlag, Berlin,
               1992.

[Zhang 92]     Z. Zhang and O.D. Faugeras. *3D Dynamic Scene Analysis*,
               Springer Verlag, 1992.

[Zhang 97]     Z. Zhang. "Parameter Estimation Techniques: A Tutorial with
               Application to Conic Fitting," *Image and Vision Computing
               Journal* Vol. 15, No. 1, pp. 59–76, 1997.

# A Multi-Ring Fiducial System and an Intensity-Invariant Detection Method for Scalable Augmented Reality

Youngkwan Cho, Jongweon Lee, and Ulrich Neumann

**Abstract.** *In Augmented Reality (AR), a user can see a virtual world as well as the real world. The user's pose in both worlds should be exactly the same to avoid the registration problem between the virtual world and the real world. Fiducial-tracking AR is an attractive approach to the registration problem; however, most of the developed fiducial-tracking AR systems have very limited tracking ranges and require carefully prepared environments, especially lighting conditions. To provide for both wide and detailed views in large-scale applications, an AR system should have a scalable tracking capability under varying light conditions.*

*In this paper, we propose multi-ring color fiducial systems and a light-invariant fiducial detection method for scalable fiducial-tracking AR systems. We analyze the optimal ring width, and develop formulas with system-specific inputs to obtain the optimal fiducial set. We present a light-invariant, circular fiducial detection method that uses relations among fiducials and their backgrounds for segmenting regions of an image. Our work provides a simple and convenient way to achieve wide-area tracking for AR.*

## 1 Introduction

In Virtual Reality, where all scenes are computer-generated images, a virtual world can be explored by flying or steering treadmills without the user's performing the same physical movements in the real world. In Augmented Reality (AR), a user sees a virtual world as well as a real world. To avoid the misalignment between two worlds, the user's pose in the real world must be directly related to the user's pose in the virtual world; AR requires the same amount of movement in both worlds. To apply AR to large-scale applications, wide-area tracking is essential.

AR's overlapped virtual world contains virtual objects to help users understand the real world [Azuma 95]. To make the virtual world useful, the virtual objects should be aligned with the real objects. Registration is one of the major issues in AR [Azuma 94] [Bajura 95] [Tuceryan 95]. The registration problem requires high accuracy or error-correction mechanisms in tracking. Fiducial tracking has been gaining interest as a solution to the registration problem [Mellor 95] [Neumann 96] [State 96] [Cho 97].

## 1.1  Motivation

The tracking range of a fiducial-tracking AR system is confined by the detectability of fiducials in input images. Most of the developed systems use their own single-size fiducials. A single-size fiducial might facilitate rapid fiducial detection, but any such system will have a narrow tracking range because all of its fiducials will have the same detection range.

In a large-scale application where details are important, an AR system should provide wide views as well as detailed (zoom-in) views of interesting regions. The developed AR systems, which use single-size fiducials, do not seem to provide such scalability. Different size fiducials will have different detection ranges. By combining a series of different detection ranges from multi-size fiducials, an extended tracking range can be created seamlessly.

Fiducial-tracking AR systems use image processing or computer vision techniques to detect fiducials; these include boundary detection, color segmentation, watershed, and template matching techniques. These methods require thresholds to be provided in order to segment an image and be able to detect fiducials; often they cannot be used under different lighting conditions without changing thresholds. Since most of the developed fiducial-tracking AR systems adopt these detection methods, they seem to require carefully prepared environments, including light conditions; therefore, their usability is very limited. To apply AR to a large-scale application, we need a detection method that works under any lighting conditions without requiring the user to control manually any parameters, including uneven lighting conditions in a single image.

## 1.2  Contribution

We add the multi-ring, multi-size concept to the concentric circular fiducials and introduce a fiducial system concept. This fiducial system concept provides great scalability for fiducial-tracking AR. The fiducial systems introduce a much larger number of unique fiducials than does a single-size fiducial system, and make fiducial identification much easier. We develop formulas to calculate the optimal fiducial set for any size application, given some system-specific parameters. Users conveniently get the optimal fiducial set simply by plugging in those parameters as dictated by their systems.

We develop a robust fiducial detection method which works under varying lighting conditions, using light-invariant relationships among homogeneous regions instead of thresholds for segmenting regions. We also develop rules and membership functions to detect fiducials.

We present a simple and low-cost way to achieve wide-area tracking; we hope that this work will trigger many research activities for large-scale applications.

## 2 Previous Works

### 2.1 Fiducial Tracking

There are several developed fiducial-tracking AR systems. Most use solid or concentric circular fiducials [Mellor 95] [Neumann 96] [State 96] [Cho 97], and a few use the corners of big rectangles as fiducials [Kutulakos 96] [Koller 97]. All current systems have adopted either single-size fiducials or rectangles; their tracking areas are limited to the periphery of a desktop or the opposite wall of a room (Figure 1). None seem to support full room ($\sim 30 \times 30$ feet) tracking from arm's-length distance to wall-to-wall distance.

### 2.2 Fiducial Detection

Several AR systems have been developed using vision-based techniques. Some authors mention in passing that their approaches deal with different

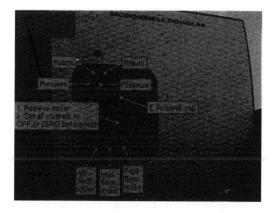

**Figure 1.** An example of an AR screen.

lighting conditions, while others do not give any information regarding the environmental restrictions under which their systems work.

Madritsch, Leberl, and Gervautz [Madritsch 96] developed a camera-based beacon tracking system. They use red LEDs as beacons and accept lighting restrictions for their system.

State, Hirota, Chen, Garrett, and Livingston [State 96] developed a hybrid system which combines a vision-based tracking system and a magnetic tracking system. Their basic algorithm for detecting fiducials is based on the ratios of RGB component values. They use the system under restricted lighting conditions and mention that the fiducial analyzer's performance diminishes with changing lighting conditions, despite the use of adaptive brightness evaluation for each landmark.

Uenohara and Kanade [Uenohara 95] mention that the appearance of objects varies in many ways, depending on pose and illumination change. They require users to locate objects at the initial recognition step. Normalized correlation and precaptured images are used to operate the system under varying lighting conditions. The complexity of the detection algorithm increases as objects become more complex and more reference images are required.

As we see from the above three systems, vision-based AR systems need a fiducial-detection algorithm; the performance of detection algorithms diminishes when lighting conditions vary. We developed a rule-based algorithm to detect fiducials with varying lighting conditions. The idea behind the algorithm is to use relations among homogeneous regions instead of using threshold values for segmenting regions.

## 3   Multi-Ring Color Fiducial System

The 2D image location $(u, v)$ of a 3D point $(x, y, z)$ is determined by the following perspective projection.

$$
\begin{bmatrix} U \\ V \\ w \end{bmatrix} = \begin{bmatrix} f_u & 0 & U_0 & 0 \\ 0 & f_v & V_0 & 0 \\ 0 & 0 & 1 & 0 \end{bmatrix} \begin{bmatrix} \mathbf{R} & \mathbf{0} \\ \mathbf{0} & 1 \end{bmatrix} \begin{bmatrix} \mathbf{I} & -\mathbf{T} \\ \mathbf{0} & 1 \end{bmatrix} \begin{bmatrix} x \\ y \\ z \\ 1 \end{bmatrix},
$$

$$
\begin{bmatrix} u \\ v \end{bmatrix} = \begin{bmatrix} U/w \\ V/w \end{bmatrix}
$$

where $f_u$ and $f_v$ are the effective focal lengths of a camera in the $u$ and $v$ direction, respectively, and $(U_0, V_0)$ is the image center. $\mathbf{R}$ is the rotation matrix, and $\mathbf{T}$ is the translation vector for the camera pose. The major axis length $d$ of the ellipse is $d = Df/w$, where $D$ is the diameter of the fiducial, $f(= f_u = f_v)$ is the effective focal length, and $w$ is the depth of the fiducial in the camera coordinate system.

When a camera is too far from or too close to a fiducial, the projected image of the fiducial in the input image will be too small or too large to detect it correctly. Therefore, an AR system with single-size fiducials has a very limited tracking range. Although each fiducial has a fixed detectable range, the whole tracking range could be extended by combining different detectable ranges of different size fiducials.

Multi-ring color fiducials have a different number of rings at different size levels. The first level fiducial has one core circle and one outer ring. As the levels go up, one extra ring is added outside of the previous level fiducial. The number of rings in a fiducial thus indicates the fiducial level. The core circle and rings are painted with six colors (red, green, blue, yellow, magenta, and cyan). They introduce many unique fiducials and make fiducial identification easier.

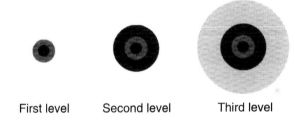

First level    Second level    Third level

**Figure 2.** Multi-ring, multi-size color fiducials.

In the proportional-width ring fiducial system, the size ratio $c$ between adjacent levels is the same.

$$D_i = cD_{i-1} \quad (c > 1)$$

$$= c^{i-1}D_1,$$

where $D_i$ is the diameter of the level $i$ fiducial. Because the outer ring is wider than the inner ring, the outer ring is more easily detectable from a distance. The higher level fiducials have a greater detectable range. By combining those detectable ranges, we can extend the whole tracking range. This fiducial system is good for wide-area tracking with arbitrary camera movements.

Let the desired tracking range be $Z_{\text{near}} \sim Z_{\text{far}}$, and the camera focal length be $f$. Let $w$ be the minimum detectable ring width in an input image. This width $w$ will depend on the camera, the digitizer, and the fiducial detection algorithm.

## 3.1 Number of Required Levels

Let the tracking range of a level $i$ fiducial be $Z_{\text{near},i} \sim Z_{\text{far},i}$, with the conditions $Z_{\text{near}} = Z_{\text{near},i}$ and $Z_{\text{far},n} = Z_{\text{far}}$. The largest detectable fiducial size in an image is $d_{\text{near}}(\geq D_i f / Z_{\text{near},i})$, and the smallest detectable fiducial size is $d_{\text{far}}(\leq D_i f / Z_{\text{far},i})$.

To combine the detectable ranges smoothly, there should be no gap between adjacent work ranges.

$$0 \leq Z_{\text{far},i} - Z_{\text{near},i+1} \leq \frac{D_i f}{d_{\text{near}}} \left( \frac{d_{\text{near}}}{d_{\text{far}}} - c \right)$$

$$c \leq \frac{d_{\text{near}}}{d_{\text{far}}}$$

The required levels of fiducials can be expressed as a function of the size ratio $c$ (Figure 3).

$$D_n = \frac{Z_{\text{far}}}{f} d_{\text{far}}, \quad D_1 = \frac{Z_{\text{near}}}{f} d_{\text{near}}$$

$$\frac{Z_{\text{far}}}{Z_{\text{near}}} \leq c^{n-1} \frac{d_{\text{near}}}{d_{\text{far}}} \geq c^n$$

$$n(c) \geq \frac{\log(Z_{\text{far}}/Z_{\text{near}})}{\log c}$$

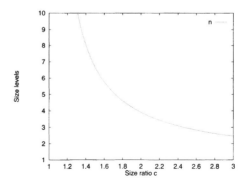

**Figure 3.** Fiducial levels.

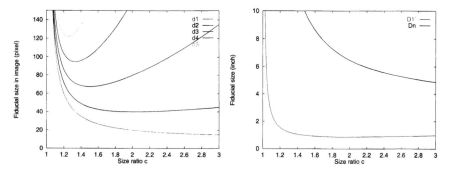

**Figure 4.** Fiducial sizes: (Left) Fiducial sizes in an image. (Right) Real fiducial sizes.

## 3.2  Fiducial Size at Each Level

When the camera is at $Z_{\text{far},1}$ from a level-$n$ fiducial, the major axis lengths of the level-1,-2, and -$j$ rings in an input image are

$$d_{\text{far},1} = \frac{2w}{c-1}c \geq d_{\text{far}}$$

$$d_2 = \frac{2w}{c-1}c^2 \leq d_{\text{near},2} \leq d_{\text{near}}$$

$$d_j = \frac{2w}{c-1}c^j, \quad (1 \leq j \leq n)$$

The diameters of the level 1 and $n$ fiducials are

$$D_n = \frac{Z_{\text{far}}}{f}d_{\text{far}} = \frac{Z_{\text{far}}}{f}\frac{2w}{c-1}c$$

$$D_1 = \frac{Z_{\text{near}}}{f}d_{\text{far}} \geq \frac{Z_{\text{far}}}{f}\frac{2w}{c-1}c^{2-n}$$

Figure 4 shows the major axis lengths of some fiducial levels, and the minimum and maximum fiducial sizes as a function of $c$.

## 3.3  Fiducial Distribution

When a camera is close to fiducials, the camera can only see a small region in the real world, and only small fiducials can be detected. When the camera is far from fiducials, it can see a large area, and only large fiducials

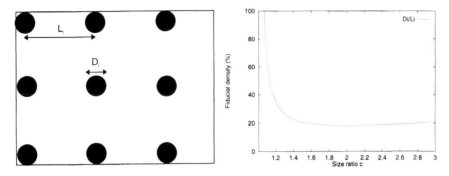

**Figure 5.** Fiducial distribution: (Left) Input image of the level $i$ fiducials at distance $Z_{\mathrm{near},i}$. (Right) Inter-fiducial distance at level 1 and $n$, and fiducial density $f(c)$.

can be detected. Therefore, lower-level fiducials have dense distributions and higher-level fiducials have sparse distributions.

Fiducials are usually distributed around interesting regions without any regular pattern. For analysis purposes, we use a regular grid distribution. We assume that the input image has $W \times H$ resolution ($W \geq H$).

To determine the camera pose, three or more non-collinear fiducials are required in an input image [Linnainmaa 88][Horaud 89][Haralick 94]. Figure 5(Left) shows an input image of level $i$ fiducials at distance $Z_{\mathrm{near},i}$. With this configuration, any camera pose can see three or more fiducials in the valid tracking range of level $i$. The inter-fiducial distance for level $i$ fiducials $L_i$ is

$$
\begin{aligned}
L_i(c) &= \frac{H - d_{\mathrm{near}}}{2} \cdot \frac{D_i}{d_{\mathrm{near}}} \\
&= \frac{H(c-1) - 2c^2 w}{4c^2 w} D_i.
\end{aligned}
$$

The ratio of the fiducial diameter over the inter-fiducial distance shows the fiducial density,

$$
f(c) = \frac{D_i}{L_i} = \frac{4c^2 w}{H(c-1) - 2c^2 w}.
$$

Figure 5(Right) shows the inter-fiducial distances and fiducial density in percentage form.

## 3.4 Optimal Size Ratio $c$

All results come out as functions of the size ratio $c$. Which $c$ gives the optimal results? Optimality has two aspects: system performance and fiducial sizes.

For high system performance, the fiducial detection process should be fast. We can concentrate on the performance of finding the smallest detectable ring in a fiducial, because the larger rings could be found easily by predicting their boundaries with $c$. The fiducial detection algorithm should look for rings whose diameters are in the range from $d_{far}(= \frac{2w}{c-1}c)$ to $d_{near}(\geq \frac{2w}{c-1}c^2)$. To minimize the processing time for fiducial detection, $f(c) = (d_{near} - d_{far})d_{far}$ should be minimized. $(d_{near} - d_{far})$ is the search range, and $d_{far}$ is the search size. The bigger $d_{far}$ is, the more processing time needed for boundary detection.

$$f(c) = d_{far}(d_{near} - d_{far}) = \frac{4w^2}{c-1}c^2$$
$$\frac{df(c)}{dc} = \frac{4w^2c(2-c)}{(c-1)^2}$$

$f(c) = (d_{near} - d_{far})d_{far}$ is at its minimum when $c = 2$.

The other aspect is fiducial size. If fiducials are too big, it is not easy to find proper places for them. They also occupy a large area in the input images. So we can calculate the optimal size by minimizing the fiducial density.

$$\frac{df(c)}{dc} = \frac{4cwH(c-2)}{(H(c-1) - 2c^2w)^2}$$

$f(c)$ has the minimum value of $16w/(H - 8w)$ when $c = 2$.

For both aspects, the optimal size ratio $c$ has the same value, 2.

## 4 Fiducial Detection

We are only interested in fiducials in an image which occupy very small areas, so we apply a multi-resolution approach to reduce the execution time of our detection algorithm. The detection algorithm is divided into two stages: coarse detection and finer detection (Figure 6). Coarse detection quickly skims through an image and finds potential regions; finer detection, the more expensive method, is applied to detect possible fiducials. Next, shape and color tests are applied to distinguish false fiducials from real

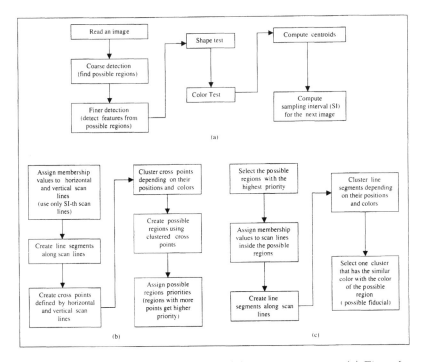

**Figure 6.** (a) Overview of the algorithm. (b) Coarse detection. (c) Finer detection.

fiducials. Coarse and finer detection methods share two steps: assigning membership values and creating line segments. They differ in two ways: coarse detection finds potential regions using only sampled horizontal and vertical scan lines, while finer detection uses all horizontal and vertical scan lines in small selected regions to detect possible fiducials; and coarse detection clusters cross points created with horizontal and vertical line segments, while finer detection clusters line segments segmented along horizontal and vertical scan lines. Clustering used in coarse and finer detection is based on locations and colors of cross points or line segments. The bases of the algorithm are rules and fuzzy algorithms. Rules are used for detecting transition areas between a fiducial and its background, and fuzzy algorithms are applied to localize the position of edges. Details of the algorithm are presented in the following sections.

## 4.1 Coarse Detection
Potential regions are detected by the coarse detection procedure. If a sampling interval is carefully designed, this procedure recognizes that cross

points are created by horizontal and vertical line segments inside a fiducial. Clusters of cross points create potential regions for a finer detection procedure.

$$s = \frac{d}{\sqrt{2}},$$

where $d$ is the size of the minor axis of the smallest fiducial. Sampling Interval (SI)$= \frac{s}{2}$.

The sampling interval is the most important parameter used in the coarse detection procedure. Small sampling intervals increase computation time, and large sampling intervals miss some fiducials. We create a potential region with a cluster that contains at least one cross point, but we use sampling intervals that create at least four cross points for each fiducial without any noise on an image. This improves robustness, even if it increases computation time. Intersections among sampled horizontal and vertical lines form square grids, and one square grid contains four cross points. If grids are contained within the inner square ($d$ equals the size of the minor axis) of the expected minimum fiducial, there are at least four cross points inside every fiducial (Figure 7). We can derive the size of the inner square of the smallest fiducial of an image from the minor axis of the smallest fiducial. Then, the optimal interval can be computed. A sampling interval changes according to the size of the minor axis of the smallest fiducial of the previous image.

For every sampled horizontal and vertical line, a membership value is assigned to each pixel. All lines are segmented according to membership values of individual pixels, and these line segments create cross points if two line segments intersect and have the same color. We cluster cross points depending on their positions and colors, and each cluster defines one

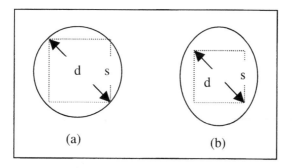

**Figure 7.** Maximum sampling interval for (a) circle, (b) ellipse.

potential region for a finer detection procedure. After eliminating clusters with a size larger than the maximum fiducial size, a priority is assigned to each potential region according to the number of cross points in a cluster.

## 4.2  Finer Detection

We next select the region with the highest priority. We assign a membership value to every pixel of all horizontal and vertical lines inside the selected region, and create line segments along all scan lines based on membership values. These line segments are clustered according to their locations and color, and clusters of line segments are possible fiducials. Sub-procedures of both detection steps are presented in following sections.

## 4.3  Rules and Membership Functions

Edge detection techniques using a fuzzy logic method usually use $S$ and $\pi$ functions as membership functions. These techniques require threshold values and prior knowledge about an input image, such as the number of regions on the image, in order to allow segmentation of an image into regions, which means that we cannot apply general membership functions to achieve our goal. We develop the membership function to find the best edge position inside the transition region. This is possible because we are looking for specific fiducials.

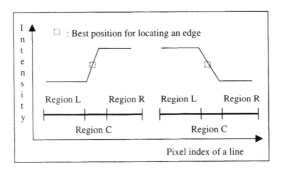

**Figure 8.** Step edge model.

The key characteristics are the minimum distance between two fiducials, the minimum and maximum size of a fiducial, and the shape of a fiducial. We also know fiducials are placed on solid backgrounds. This means that interesting edges are located between two homogeneous regions (Figure 8). We can extract a few rules and membership functions from the relations between homogeneous regions. These rules eliminate nearly all false edges in an input image, and membership functions locate the best positions for edges. The best location for an edge is a pixel location whose intensity is the same as the average value of left and right segments of an edge. Rules and membership functions are:

1. Monotonic average rule:

   $\text{Avg}(R) > \text{Avg}(C) > \text{Avg}(L)$ or $\text{Avg}(L) > \text{Avg}(C) > \text{Avg}(R)$, where $\text{Avg}(J)$ is an average intensity value of a region $J$.

2. Distribution rule:

   $\max(C) - \min(C) > \max(R) - \min(R)$ and $\max(C) - \min(C) > \max(L) - \min(L)$, where $\max(J)$ and $\min(J)$ indicate the maximum and minimum intensity values of a region $J$.

3. Overlapping rule:

   $\max(R) < \min(L)$ when $\text{Avg}(R) < \text{Avg}(L)$
   $\max(L) < \min(R)$ when $\text{Avg}(L) < \text{Avg}(R)$

4. Membership functions:

$$\mu = \mu 1 \times \mu 2$$

$$\mu 1 = \frac{2 \times \min\left(|Avg(R) - Avg(C)|, |Avg(L) - Avg(C)|\right)}{|Avg(R) - Avg(L)|}$$

$$\mu 2 = \frac{\min(R) - \max(L)}{\max(R) - \min(L)}, Avg(R) > Avg(L)$$

*or*

$$\mu 2 = \frac{\min(L) - \max(R)}{\max(L) - \min(R)}, Avg(L) > Avg(R)$$

The membership function contains two parts, $\mu 1$ and $\mu 2$. $\mu 1$ indicates the grade of closeness to the median intensity value between two regions. $\mu 1 = 1$ if $A(C) = |A(R) - A(L)|/2$, and $\mu 1 < 1$ for other cases. $\mu 2$ indicates the grade of closeness to the ideal edge. If an edge is the ideal edge, $(\min(R) - \max(L)) = (\max(R) - \min(L))$ or $(\min(L) - \max(R)) = (\max(L) - \min(R))$, and $\mu 2 = 1$. $\mu 2 < 1$ for other cases. Therefore the membership function $\mu$ is used to find the position of an edge that is close to the ideal edge.

## 4.4   Detect Line Segments

Next we collect pixels from each horizontal scan line that passes all the rules
listed above. We group these pixels, which are connected without crossing
a pixel that does not pass all the rules. From each group, we select one
pixel with the highest membership value, which indicates the best location
of an edge. We connect the selected pixels to create line segments.

## 4.5   Cluster Line Segments

Each fiducial has a solid color, so we can find a position of a fiducial by
clustering the line segments with the same color. Unfortunately, every image contains noise. For example, a region with solid color has pixels with
different color values. Thus, we must introduce a similarity measure. Traditionally, similarity is measured with a distance metric and a threshold.
Possible distance metrics used for color similarity include absolute distance
(e.g., Manhattan distance, Euclid distance), 1-norm distance, 2-norm distance, $\infty$-norm distance, angle between colors in the RGB color cube, and
the square of the cosine of the color angle.

   These metrics require thresholds in order to enable users to decide
whether two color values are similar. Defining a threshold that works for
varying lighting conditions is difficult, because color values change with
lighting conditions. Therefore, we developed a similarity measure that
uses the probability theory and utilizes local information existing on line
segments. A uniform probability density function is created for each line
segment, and two line segments are considered to have the same color when
two uniform probability density functions overlap.

   This is possible because color similarity is checked when two line segments are next to each other, and fiducials and their background have different colors. The distribution of a region is defined by the minimum, average,
and maximum values of a region. We find $D$, $Min(|A(J)-\min(J)|, |A(J)-\max(J)|)$, and create a uniform distribution by $A(J)D$ and $A(J) + D$,
where $Min(A, B)$ is a smaller value between $A$ and $B$. Since we choose
$Min(|A(J) - \min(J)|, |A(J) - \max(J)|)$ as $D$, effects of noise pixels can
be eliminated; $D$ is used to define a uniform density function representing
an intensity distribution of a region. This simple density function works
well for our detection algorithm, and it reduces the computation time of
clustering procedures.

# 5   Result and Discussion

Our implementation has the following configuration:

- SGI Indy 24-bit graphics system with 4400 MIPS at 200MHz.

- SONY DXC-151A color video camera with $640 \times 480$ resolution, $31.4°$ in horizontal and $24.37°$ in vertical FOV, S-video output.

- A three-level proportional fiducial set with six colors (red, green, blue, yellow, cyan, and magenta). The diameter of the first-level fiducials is 1", the second-level 2", and the third-level 4".

The smallest detectable ring width of our implementation is 7 pixels. We search rings with 24 to 56 pixels in diameter. The detection range of the first-level fiducial is 1.7' to 3.9', the second-level 3.3' to 7.7', the third-level 6.6' to 15.4' (Table 1.). Therefore, the whole detection range is 1.7' to 15.4'. Figure 9 shows three snapshots of detection results from typical distances for the three levels under uneven lighting conditions. The detected fiducials are marked with white cross hairs at the centers.

The system performance depends on the number and size of the potential fiducials in an image. The current implementation does not use any prediction of fiducial positions, but skims the whole image every time. Even a linear prediction could improve the system performance by reducing search areas for coarse detection.

As Figure 9 (Left) shows, large fiducials can be detected at close distance. Although a whole fiducial is too large in the image, one of the small rings in the fiducial might be detected. After finding one ring, we can predict and find outer rings easily because the fiducial system uses a proportional ring width. Eventually the whole fiducial can be detected. Therefore, in the multi-ring fiducial system, large fiducials can be used at

**Figure 9.** Detection results: The detected fiducials have white cross hairs at the center. (Left) Distance 3 feet — All three level fiducials are detected. (Center) Distance 6 feet — The second- and third-level fiducials are detected. (Right) Distance 12 feet — Only the third-level fiducials are detected.

| Fiducial level | Diameter (inch) | Theoretical tracking range(feet) | Snapshot distance (feet) | Frame rate (FPS) |
|---|---|---|---|---|
| First level | 1 | 1.7-3.9 | 3,(a) | 1.0 |
| Second level | 2 | 3.3-7.7 | 6,(b) | 1.5 |
| Third level | 4 | 6.6-15.4 | 12,(c) | 2.0 |

**Table 1.** Detection results.

close range. The partially visible large fiducials can also be detected. These are the advantages of the multi-ring fiducial system.

We use two lighting sources (daylight and fluorescent light) as well as different apertures of a camera to simulate different lighting conditions. The algorithm is tested on images with different backgrounds and apertures, $f = 1.8$ to $8.0$. The backgrounds of fiducials are small regions around fiducials (Figure 10). The algorithm is also tested on live video sequence after it is integrated into the existing AR system. Our detection method is compared with the gradient-based detection method used for the current AR system [Cho 97]. The gradient-based detection method detects most fiducials correctly with $f = 2.0$ to $4.0$. The presented algorithm detects all fiducials under every aperture setting except yellow fiducials on a white background at $f = 1.8$ and green fiducials on a black background at $f = 8.0$ and $5.6$ (Table 2). Undetectable fiducials are not easily perceived by the human eye.

This fiducial detection algorithm is robust with varying lighting conditions, and it is unique in that it uses rules and membership functions extracted from relations among fiducials and homogeneous backgrounds. The algorithm detects fiducials under varying lighting conditions, without the need for any human intervention; it does so including uneven lighting

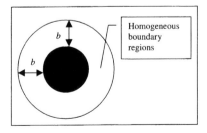

**Figure 10.** Defining the background of a fiducial: $b$ is the size of a homogeneous boundary region considered the background of a fiducial.

| Number of Detected Fiducials | | 11 | 12 | 12 | 12 | 12 | 11 | 11 | 12 | 12 | 12 | 11 | 11 |
|---|---|---|---|---|---|---|---|---|---|---|---|---|---|
| White Background | R | Y | Y | Y | Y | Y | Y | Y | Y | Y | Y | Y | Y |
| | G | Y | Y | Y | Y | Y | Y | Y | Y | Y | Y | Y | Y |
| | B | Y | Y | Y | Y | Y | Y | Y | Y | Y | Y | Y | Y |
| | Y | N | Y | Y | Y | Y | Y | N | Y | Y | Y | Y | Y |
| | C | Y | Y | Y | Y | Y | Y | Y | Y | Y | Y | Y | Y |
| | M | Y | Y | Y | Y | Y | Y | Y | Y | Y | Y | Y | Y |
| Black Background | R | Y | Y | Y | Y | Y | Y | Y | Y | Y | Y | Y | Y |
| | G | Y | Y | Y | Y | Y | N | Y | Y | Y | Y | N | N |
| | B | Y | Y | Y | Y | Y | Y | Y | Y | Y | Y | Y | Y |
| | Y | Y | Y | Y | Y | Y | Y | Y | Y | Y | Y | Y | Y |
| | C | Y | Y | Y | Y | Y | Y | Y | Y | Y | Y | Y | Y |
| | M | Y | Y | Y | Y | Y | Y | Y | Y | Y | Y | Y | Y |
| Aperture (f) | | 1.8 | 2.0 | 2.8 | 4.0 | 5.6 | 8.0 | 1.8 | 2.0 | 2.8 | 4.0 | 5.6 | 8.0 |
| Lighting source | | Day light | | | | | | Fluorescent light | | | | | |
| R: Red, G: Green, B: Blue, Y: Yellow, C: Cyan, and M: Magenta | | | | | | | | | | | | | |

**Table 2.** Results of fiducial detection algorithm. (Y indicates detection and N indicates no detection.)

conditions in a single image. The algorithm thus extends the usability of vision-based AR systems quite a bit, since an AR system with the presented algorithm can be used under varying lighting conditions.

Fiducials are distributed on the interested objects and/or environments in a real world. The number of fiducials required in the real world depends on the application size. For full-room (30' × 30') tracking, there might be tens or hundreds of fiducials in the room. Because of the FOV of a camera, the camera can only see part of a room, and each input image will contain only a few fiducials. To determine the camera pose, correspondence between the real fiducials and the image fiducials should be established. This is our future research direction.

## Acknowledgments

We acknowledge support by NSF grant No. CCR-9502830 and the USC Integrated Media Systems Center, an NSF ERC. We thank Anthony Majoros and McDonnell Douglas Aerospace for their invaluable assistance in defining the applications, and for the loan of the aircraft section model shown in Figure 1.

## References

[Azuma 94]    R. Azuma and G. Bishop. "Improving Static and Dynamic Registration in an Optical See-Through HMD," *Proceedings of 21st International SIGGRAPH Conference,* ACM, Orlando, FL, pp. 197–204, 1994.

[Azuma 95]        R. Azuma. "A Survey of Augmented Reality." *SIGGRAPH 1995, Course Notes #9*, Los Angeles, CA, 1995.

[Bajura 95]       M. Bajura and U. Neumann. "Dynamic Registration Correction in Augmented Reality Systems," *Proccedings of Virtual Reality Annual International Symposium (VRAIS)*, Research Triangle Park, NC, pp. 189–196, 1995.

[Cho 97]          Y. Cho, J. Park, and U. Neumann. "Fast Color Fiducial Detection and Dynamic Workspace Extension in Video See-through Augmented Reality," *Proceedings of the Fifth Pacific Conference on Graphics and Applications,* pp. 168–177, October 1997.

[Haralick 94]     R. Haralick, C. Lee, K. Ottenberg, and M. Nolle. "Review and Analysis of Solutions of the Three Point Perspective Pose Estimation Problem," *IJCV,* Vol. 13, No. 3, Seoul, Korea, pp. 331–356, 1994.

[Horaud 89]       R. Horaud, B. Conio, and O. Leboulleux. "An Analytic Solution for the Perspective 4-Point Problem," *CVGIP* 47, pp. 33–44, 1989.

[Koller 97]       D. Koller, G. Klinker, E. Rose, D. Breen, R. Hitaker, and M. Tuceryan. "Real-time Vision-Based Camera Tracking for Augmented Reality Applications," *Proceedings of the Symposium on VRST,* Lausanne, Switzerland, pp. 87–94, September, 1997.

[Kutulakos 96]    K.N. Kutulakos and J. Vallino. "Affine Object Representations for Calibration-free Augmented Reality," *Proccedings of Virtual Reality Annual International Symposium (VRAIS)* Santa Clara, CA, pp. 25–36, 1996.

[Linnainmaa 88]   S. Linnainmaa, D. Harwood, and L.S. Davis. "Pose Determination of a Three-Dimensional Object Using Triangle Pairs," *PAMI,* Vol. 10, No. 5, pp. 634–647, September 1988.

[Madritsch 96]    F. Madritsch, F. Leberl, and M. Gervautz. "Camera Based Beacon Tracking: Accuracy and Applications," *Proceedings of ACM Symposium on Virtual Reality Software and Technology (VRST 96),* Hong Kong, pp. 101–108, 1996.

[Mellor 95]       J.P. Mellor. "Enhanced Reality Visualization in a Surgical Environment," *AI Technical Report 1544,* 1995.

[Neumann 96]      U. Neumann and Y. Cho. "A Self-Tracking Augmented Reality System," *Proceedings of ACM Symposium on Virtual Reality Software and Technology (VRST '96),* pp. 109–115, Hong Kong, July 1996.

[State 96]        A. State, G. Hirota, D.T. Chen, B. Garrett, and M. Livingston. "Superior Augmented Reality Registration by Integrating Landmark Tracking and Magnetic Tracking," *SIGGRAPH 1996,* New Orleans, LA, pp. 429–438, 1996.

[Tuceryan 95]    M. Tuceryan, D.S. Greer, P.T. Whitaker, D. Breen, C. Cramp-
ton, E. Rose, and K.H. Ahlers. "Calibration Requirements and
Procedures for a Monitor-Based Augmented Reality System,"
*IEEE Transactions on Visualization and Computer Graphics,*
Vol. 1, No. 3, pp. 255–273, September 1995.

[Uenohara 95]    M. Uenohara and T. Kanade. "Vision-Based Object Registration for Real-Time Image Overlay," *Proceedings of Computer
Vision, Virtual Reality, and Robotics in Medicine 95,* Nice, pp.
13–22, 1995.

# Fusion of Data from Head-Mounted and Fixed Sensors

William A. Hoff

**Abstract.** *A methodology is developed for explicitly fusing sensor data from a combination of fixed and head-mounted sensors, in order to improve the registration of objects in an augmented reality system. The methodology was applied to the analysis of an actual experimental augmented reality system, incorporating an optical see-through, head-mounted display; a head-mounted CCD camera; and a fixed optical tracking sensor. The purpose of the sensing system was to determine the position and orientation (pose) of a movable object with respect to the head-mounted display. A typical configuration was analyzed, and it was shown that the hybrid system produces a pose estimate that is significantly more accurate than that produced by either sensor acting alone. When only the fixed sensor was used, the maximum translational error in the location of an object in any direction from the head-mounted display was 8.23 mm (corresponding to a 97% confidence interval). When only the head-mounted sensor was used, the maximum translational error in any direction was 19.9 mm. When data from the two sensors was combined, the maximum translational error was reduced to 1.47 mm. In order to fuse the pose estimates, the uncertainties are explicitly calculated, in the form of covariance matrices. A capability was also developed to visualize the uncertainties as three-dimensional ellipsoids.*

## 1   Introduction

Where is an object of interest with respect to the user's head? In augmented reality systems that use head-mounted displays (HMDs), knowing the relative position and orientation (pose) between object and head is crucial for displaying a virtual object that is aligned with the real object. If the estimated pose of the object is inaccurate, the real and virtual objects may not be registered correctly. Registration inaccuracy is one of the most important problems limiting augmented reality applications [Azuma 97].

167

To determine the pose of an object with respect to the user's head, tracking sensors are necessary. Optical sensors use cameras or photo-effect sensors to track optical targets, such as light emitting diodes (LEDs) or passive fiducial markings [Azuma 94] [Wang 90] [Kim 97]. If one uses two or more sensors (stereo vision), the three-dimensional (3D) position of a target point can be determined directly via triangulation. The accuracy of locating the point is improved by increasing the separation (baseline) between the sensors. The full six degree-of-freedom (DOF) pose of a rigid body can be determined by measuring three or more target points on the body, assuming that the geometry of the points on the body is known. In the photogrammetry literature this procedure is known as the "absolute orientation" problem. Alternatively, a single sensor can be used to measure the 2D (image) locations of three or more target points on a rigid body. If the geometry of the points is known, the full six DOF pose of the rigid body can be estimated, by a procedure known as "exterior orientation" [Haralick 93].

One issue that arises is where to place the sensor and target. One possibility is to mount the sensor at a fixed known location in the environment, and put targets on both the HMD and the object of interest (a configuration called "outside in" [Wang 90]). We then measure the pose of the HMD with respect to the sensor and the pose of the object with respect to the sensor, and derive the relative pose of the object with respect to the HMD. Another possibility is to mount the sensor on the HMD, and the target on the object of interest (a configuration called "inside out"). We measure the pose of the object with respect to the sensor, and use the known sensor-to-HMD pose to derive the relative pose of the object with respect to the HMD. Both approaches have been tried in the past, and each has advantages and disadvantages.

With a fixed-sensor (outside-in approach) there is no limitation on size and weight of the sensor. Multiple cameras can be used, with a large baseline, to achieve highly accurate 3D measurements via triangulation. For example, commercial optical measurement systems such as Northern Digital's Optotrak have baselines of approximately one meter and are able to measure the 3D positions of LED markers to an accuracy of less than 0.25 mm. The orientation and position of a target pattern is then derived from the individual point positions. A disadvantage of this approach is that head orientation must be inferred indirectly from the point positions.

The inside-out approach has good registration accuracy, because a slight rotation of a head-mounted camera causes a large shift of a fixed target in the image. However, a disadvantage of this approach is that it is impossible to put multiple cameras with a large baseline separation on the head.

Either a small baseline separation or a single camera must be used with the exterior orientation algorithm. Either method gives rise to large translation errors along the line of sight of the cameras.

A question arises — is it possible to fuse the data from a head-mounted sensor and a fixed sensor in order to derive a more accurate estimate of object-to-HMD pose? If the data from these two types of sensors are complementary, then the resulting pose can be much more accurate than one derived from any sensor used alone. We can effectively create a hybrid system that combines the inside-out and outside-in approaches.

This paper describes a methodology for explicitly computing uncertainties of pose estimates, propagating these uncertainties from one coordinate system to another, and fusing pose estimates from multiple sensors. The contribution of this work is the application of this methodology to the registration problem in augmented reality. It is shown that a hybrid sensing system, combining both head-mounted and fixed sensors, can improve registration accuracy.

## 2 Background on Pose Estimation

### 2.1 Representation of Pose

The notation in this section follows that of Craig [Craig 90]. The pose of a rigid body $A$ with respect to another coordinate system $B$ can be represented by a six-element vector

$$
{}^B_A X = ({}^B x_{Aorg}, {}^B y_{Aorg}, {}^B z_{Aorg}, \alpha, \beta, \gamma)^T,
$$

where ${}^B P_{Aorg} = ({}^B x_{Aorg}, {}^B y_{Aorg}, {}^B z_{Aorg})^T$ is the origin of frame $A$ in frame $B$, and $(\alpha, \beta, \gamma)$ are the angles of rotation of $A$ about the $(z, y, x)$ axes of $B$. An alternative representation of orientation is to use three elements of a quaternion; the conversion between $xyz$ angles and quaternions is straightforward. Equivalently, pose can be represented by a $4 \times 4$ homogeneous transformation matrix:

$$
{}^B_A H = \begin{pmatrix} {}^B_A R & {}^B P_{Aorg} \\ 0 & 1 \end{pmatrix} \tag{1}
$$

where ${}^B_A R$ is a $3 \times 3$ rotation matrix. In this paper, we shall use the letter $X$ to designate a six-element pose vector and the letter $H$ to designate the equivalent $4 \times 4$ homogeneous transformation matrix.

Homogeneous transformations are a convenient and elegant representation. Any homogeneous point $^AP = (^Ax_p, ^Ay_p, ^Az_p, 1)^T$ represented in coordinate system $A$ may be transformed to coordinate system $B$ with a simple matrix multiplication $^BP = ^B_A H ^AP$. The homogeneous matrix representing the pose of frame $B$ with respect to frame $A$ is just the inverse of the pose of $A$ with respect to $B$; i.e., $^A_B H = ^B_A H^{-1}$ . Finally, if we know the pose of $A$ with respect to $B$, and the pose of $B$ with respect to $C$, then the pose of $A$ with respect to $C$ is easily given by the matrix multiplication $^C_A H = ^C_B H ^B_A H$.

## 2.2   Pose Estimation Algorithms

The problem of determining the pose of a rigid body, given an image from a single camera, is called the "exterior orientation" problem in photogrammetry. Specifically, we are given a set of known 3D points on the object (in the coordinate frame of the object), and the corresponding set of 2D measured image points from the camera, which are the perspective projections of the 3D points. The internal parameters of the camera (focal length, principal point, etc.) are known. The goal is to find the pose of the object with respect to the camera, $^{cam}_{obj} X$. There are many solutions to the problem; in this work we used the algorithm described by Haralick [Haralick 93], which uses an iterative, non-linear, least squares method. The algorithm effectively minimizes the squared error between the measured 2D point locations and the predicted 2D point locations.

The problem of determining the pose of a rigid body, given a set of 3D point measurements, is called the "absolute orientation problem" in photogrammetry. These 3D point measurements may have been obtained from a previous triangulation process, using a sensor consisting of multiple cameras. Specifically, we are given a set of 3D known points on the object $^{obj}P_i$, and the corresponding set of 3D measured points from the sensor $^{sen}P_i$. The goal is to find the pose of the object with respect to the sensor, $^{sen}_{obj} X$. There are many solutions to the problem; in this work we used the algorithm described by [Horn 87] which uses a quaternion-based method.

## 3   Determination and Manipulation of Pose Uncertainty

Once we have estimated the pose of an object using one of the methods above, what is the uncertainty of the pose estimate? Knowing the uncertainty is critical to fusing measurements from multiple sensors. We can represent the uncertainty of a six-element pose vector $X$, by a $6 \times 6$ covariance matrix $C_X = E(\Delta X \ \Delta X^T)$ , which is the expectation of the square of the difference between the estimate and the true vector. This section describes methods for estimating the covariance matrix of a pose, transforming the covariance matrix from one coordinate frame to another, and combining two pose estimates.

## 3.1 Computation of Covariance

Assume we have $N$ measured data points from the sensor $(P_1, P_2, \ldots, P_N)$, and the corresponding points on the object $(Q_1, Q_2, \ldots, Q_N)$. The object points $Q_i$ are 3D; the data points $P_i$ are either 3D (in the case of 3D-to-3D pose estimation) or 2D (in the case of 2D-to-3D pose estimation). We assume that the noise in each measured data point is independent, and the noise distribution of each point is given by a covariance matrix $C_P$.

Let $P_i = H(Q_i, X)$ be the function which transforms object points into data points. In the case of 3D-to-3D pose estimation, this is just a multiplication of $Q_i$ by the corresponding homogeneous transformation matrix. In the case of 2D-to-3D pose estimation, the function is composed of a transformation followed by a perspective projection. An algorithm that solves for $X_{est}$ minimizes the sum of the squared errors. Assume that have we solved for $X_{est}$ using the appropriate algorithm (i.e., 2D-to-3D or 3D-to-3D). We then linearize the equation about the estimated solution $X_{est}$:

$$P_i + \Delta P_i = H(Q_i, X_{est} + \Delta X) \approx H(Q_i, X_{est}) + \left[\frac{\partial H}{\partial X}\right]^T_{Q_i, X_{est}} \Delta X \quad (2)$$

Since $P_i = H(Q_i, X_{est})$, the equation reduces to

$$\Delta P_i = \left[\frac{\partial H}{\partial X}\right]^T_{Q_i, X_{est}} \Delta X = M_i \Delta X \quad (3)$$

where $M_i$ is the Jacobian of $H$, evaluated at $(Q_i, X_{est})$. Combining all the measurement equations, we get the matrix equation:

$$\begin{pmatrix} \Delta P_1 \\ \vdots \\ \Delta P_N \end{pmatrix} = \begin{pmatrix} M_1 \\ \vdots \\ M_N \end{pmatrix} \Delta X \Rightarrow \Delta P = M \Delta X \quad (4)$$

Solving for $\Delta X$, we get $\Delta X = (M^T M)^{-1} M^T \Delta P$. The covariance matrix of $X$ is given by the expectation of the outer product:

$$
\begin{aligned}
C_x &= E(\Delta X \Delta X^T) \\
&= E\left[ (M^T M)^{-1} M^T \Delta P \Delta P^T \left( (M^T M)^{-1} M^T \right)^T \right] \\
&= (M^T M)^{-1} M^T E(\Delta P \Delta P^T) \left( (M^T M)^{-1} M^T \right)^T \\
&= (M^T M)^{-1} M^T
\begin{pmatrix}
C_P & \cdots & 0 \\
\vdots & \ddots & \vdots \\
0 & \cdots & C_P
\end{pmatrix}
\left( (M^T M)^{-1} M^T \right)^T \quad (5)
\end{aligned}
$$

Note that we have assumed that the errors in the data points are independent; i.e., $E(\Delta P_i \Delta P_j^T) = 0$, for $i \neq j$. (If the errors in different data points are actually correlated, our simplified assumption could result in an underestimate of the actual covariance matrix.) The above analysis was verified with Monte Carlo simulations, using both the 3D-to-3D algorithm and the 2D-to-3D algorithm.

## 3.2   Interpretation of Covariance

A useful interpretation of the covariance matrix can be obtained by assuming that the errors are jointly Gaussian. The joint probability density for $N$-dimensional error vector $\Delta X$ is [Trees 68]:

$$
p(\Delta X) = \left( |2\pi|^{N/2} |C_X|^{1/2} \right)^{-1} \exp\left( -\frac{1}{2} \Delta X^T C_X^{-1} \Delta X \right) \quad (6)
$$

If we look at surfaces of constant probability, the argument of the exponent is a constant, given by the relation $\Delta X^T C_X^{-1} \Delta X = z^2$ . This is the equation of an ellipsoid in $N$ dimensions. For a given value of $z$, the cumulative probability of an error vector being inside the ellipsoid is $P$. For $N=3$ dimensions, the ellipsoid defined by $z=3$ corresponds to a cumulative probability of approximately 97%.[1]

For a six-dimensional pose $X$, the covariance matrix $C_X$ is $6 \times 6$, and the corresponding ellipsoid is six-dimensional (which is difficult to visualize). However, we can select only the 3D translational component of the pose and look at the covariance matrix corresponding to it. Specifically, let $Z = (x, y, z)^T$ be the translational portion of the pose vector $X = (x, y, z, \alpha, \beta, \gamma)^T$. We obtain $Z$ from $X$ using the equation $Z = MX$,

---

[1]The exact formula for the cumulative probability in $N$ dimensions is $1 - P = \frac{N}{2^{N/2}\Gamma(N/2+1)} \int_z^\infty X^{N-1} e^{-X^2/2} dX$ [Trees 68]

where $M$ is the matrix

$$M = \begin{pmatrix} 1 & 0 & 0 & 0 & 0 & 0 \\ 0 & 1 & 0 & 0 & 0 & 0 \\ 0 & 0 & 1 & 0 & 0 & 0 \end{pmatrix}. \tag{7}$$

The covariance matrix for $Z$ is given by $C_Z = MC_X M^T$ (which is just the upper left $3 \times 3$ submatrix of $C_X$). We can then visualize the three-dimensional ellipsoid corresponding to $C_Z$.

## 3.3 Transformation of Covariance

We can transform a covariance matrix from one coordinate frame to another. Assume that we have a six-element pose vector $X$ and its associated covariance matrix $C_X$. Assume that we apply a transformation, represented by a six-element vector $W$, to $X$ to create a new pose $Y$. Denote $Y = g(X, W)$. A Taylor series expansion yields $\Delta Y = J \Delta X$, where $J = (\partial g / \partial X)$. The covariance matrix $C_Y$ is found by:

$$C_Y = E(\Delta Y \Delta Y^T) = E\left[(J\Delta X)(J\Delta X)^T\right] = JE(\Delta X \Delta X^T)J^T = JC_X J^T. \tag{8}$$

A variation on this method is to assume that the transformation $W$ also has an associated covariance matrix $C_W$. In this case, the covariance matrix $C_Y$ is:

$$C_Y = J_X C_X J_X^T + J_W C_W J_W^T \tag{9}$$

where $J_X = (\partial g / \partial X)$ and $J_W = (\partial g / \partial W)$. The above analysis was verified with Monte Carlo simulations, using both the 3D-to-3D algorithm and the 2D-to-3D algorithm.

## 3.4 Combining Pose Estimates

Two vector quantities may be fused by averaging them, weighted by their covariance matrices. Let $X_1$, $X_2$ be two $N$-dimensional vectors, and $C_1$, $C_2$ be their $N \times N$ covariance matrices. Assuming $X_1$ and $X_2$ are uncorrelated, then the combined estimate $X$ and the combined covariance matrix $C$ may be found by the following equations[2]:

$$\begin{aligned} X &= C_2(C_1 + C_2)^{-1}X_1 + C_1(C_1 + C_2)^{-1}X_2 \\ C &= C_2(C_1 + C_2)^{-1}C_1. \end{aligned} \tag{10}$$

---

[2]These equations can be derived from the discrete Kalman filter update equations, using $X_1$ as the *a priori* estimate, $X_2$ as the measurement, and $X$ as the *a posteriori* estimate.

This is a method of sensor fusion in the hybrid augmented reality system. If the pose of an object with respect to the HMD can be estimated using data from the head-mounted sensor, and the same pose can be estimated using data from the fixed sensor, then a combined estimate can be produced using Equation 10.

When combining pose estimates, we use a quaternion-based representation of orientation, rather than $xyz$ angles or Euler angles. This is because $xyz$ angles pose a problem for orientations where one angle is close to 180°: in this case, one of the pose vectors may have a value for the angle close to $+180°$, and the other vector may have a value close to $-180°$. Even though the two vectors represent very similar orientations, the combined vector would represent a wildly different orientation. Quaternions do not have this problem.

# 4    Experiments

The methodology described in the previous sections was applied to an actual experimental augmented reality system developed in our lab. The purpose of the system is to display a graphical overlay on an HMD, such that the overlay is registered to a movable object in the scene. Only quasi-static registration is considered in this paper; that is, objects are stationary when viewed, but can be moved freely. The system incorporates both head-mounted and fixed sensors. The hybrid system was developed for a surgical aid application, but its capabilities are such that it could be used in many other applications.

The first sub-section below describes the experimental set-up, including the sensor and display characteristics. Additional details of the system are described in Hoff et al.[Hoff 96]. In the second sub-section, the task itself is described. Finally, an analysis of registration accuracy is performed.

## 4.1    Description of Experimental AR System

The prototype augmented reality system incorporates a see-through HMD (Virtual I-O i-glasses$^{TM}$) mounted on a helmet (Figure 1, left). A CCD camera with a field of view of 44 degrees is also mounted on the helmet. The NTSC-format video signal from the camera is transmitted to a PC through a cable tether, which digitizes and processes the image. An optical target is affixed to the object of interest (Figure 1, right). For this work, we used a pattern of 5 green LEDs, in a rectangular planar configuration. The distinctive geometric pattern of the LEDs enables the correspondence to be easily determined [Hoff 96].

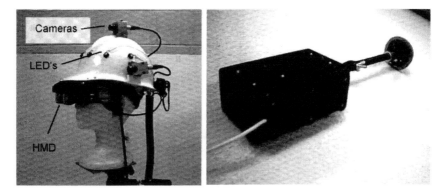

**Figure 1.** (Left) Prototype augmented reality system. Only one of the cameras was used. (Right) Five green LED's (top surface) form the optical target for the head mounted camera. Six infrared LED's (front surface) form an optical target for the Optotrak sensor. Both targets are mounted on a box, which is attached to a hip implant component.

The PC performs low-level image processing to extract the image locations of the LED targets. The noise in the 2D measured image point locations was assumed to be isotropic, with an estimated standard deviation of 0.5 pixels. Pose estimation is done using a 2D-to-3D algorithm. The throughput currently achieved with the system is approximately 8.3 Hz.

Our fixed sensor was an optical measurement system (Northern Digital Optotrak 3020) fastened to one wall of the laboratory. The sensor consists of three linear-array CCD cameras. An optical target, consisting of a set of six infrared LEDs, is fastened to each object of interest. The cameras detect each LED and calculate (via triangulation) its 3D location with respect to the sensor. From the resulting set of 3D point positions on a target body, the controller also calculates the pose of the body with respect to the sensor. For 18 target points, we measured an update rate of approximately 4 Hz.

Infrared LEDs were also placed on the helmet to form an optical target. A set of 6 LEDs were mounted in a semi-circular ring around the front half of the helmet (Figure 1, left). Typically, only 4 LEDs were visible at any one time. The measurement noise was assumed to be isotropic, with $\sigma = 0.15$ mm.

## 4.2 Description of Task

The hybrid augmented reality system was developed for a surgical aid application—specifically, total hip joint replacement. The purpose of the augmented reality system is to track a hip implant and display a graphical

| Frame | Description |
|---|---|
| HMD | Centered at left eyepiece of display |
| Implant | Centered on implant component |
| HMD target | Optical target mounted on helmet, tracked by fixed sensor |
| Camera | Camera mounted on helmet |
| Implant target | Optical target attached to implant, tracked by fixed sensor |
| Camera target | Optical target attached to implant, tracked by head-mounted camera |

**Table 1.** Principal coordinate frames in the system.

overlay on the HMD that is registered to the implant. Optical targets were attached to the implant to enable sensor tracking (shown in Figure 1, right). Separate LED targets were used for the head-mounted and fixed sensors.

The principal coordinate frames used in the system are listed and described in Table 1, and depicted schematically in Figure 2. Although this figure shows all frames as co-planar, the transformations between frames are actually fully six-dimensional (*i.e.*, three translational and three rotational components).

To aid in visualizing these coordinate frames, a 3D graphical display system was developed using a Silicon Graphics workstation and the "Open

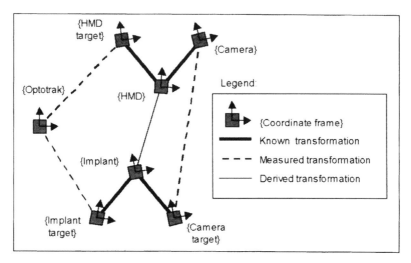

**Figure 2.** The principal coordinate frames in the system are shown, along with the transformations between them.

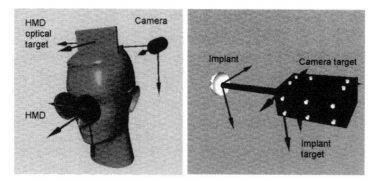

**Figure 3.** The coordinate frames on the head (left) and on the implant (right).

Inventor" graphics package. Figure 3 (left) shows a simplified representation of the coordinate frames on the head: the HMD, the HMD target, and the head-mounted camera. These coordinate frames are rigidly mounted with respect to each other on the helmet. Figure 3 (right) shows a simplified representation of the coordinate frames attached to the implant: the implant, the implant target, and the camera target. These coordinate frames are also rigidly mounted with respect to each other. (The real helmet and implant assemblies were shown in Figure 1.) The coordinate axes of all frames are also shown.

Figure 4 (left) shows the entire room scene, consisting of the fixed sensor on the back wall, the observer with the HMD, and the patient on the table with the hip implant. Figure 4 (right) shows a 3D visualization of the same scene.

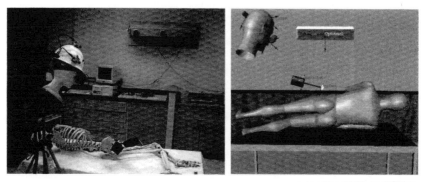

**Figure 4.** A visualization of the entire scene, showing the fixed sensor on the wall, the HMD, and the hip implant. (Left) The real scene. (Right) A 3D visualization.

## 4.3   Analysis of Registration Accuracy

A simulation was implemented, using the software application *Mathematica*, to estimate the accuracy of the derived implant-to-HMD pose. The processing consists of three main steps. First, an estimate of implant-to-HMD pose is derived using data obtained from the Optotrak (fixed) sensor alone. Second, an estimate of implant-to-HMD pose is derived using data obtained from the head-mounted camera alone. Finally, the two estimates are fused to produce a single, more accurate estimate. These steps are described in detail below.

### Pose Estimation from Fixed Sensor

Using data from the fixed sensor (Optotrak), we estimated the pose of the HMD target ($^{\text{Optotrak}}_{\text{HmdTarg}}H$) with respect to the sensor, using the 3D-to-3D algorithm described earlier. From the estimated error in each 3D point measurement (0.15 mm), the covariance matrix of the resulting pose was determined. Using the known pose of the HMD with respect to the HMD target ($^{\text{HmdTarg}}_{\text{Hmd}}H$), the pose of the HMD with respect to the sensor was estimated, using the equation

$$^{\text{Optotrak}}_{\text{Hmd}}H = {}^{\text{Optotrak}}_{\text{HmdTarg}}H \; {}^{\text{HmdTarg}}_{\text{Hmd}}H.$$

The covariance matrix of the resulting pose was also estimated. The ellipsoids corresponding to the uncertainties in the translational components of the poses are shown in Figure 5, left. In all figures, the ellipsoids are drawn corresponding to a normalized distance of $z = 3$; i.e., corresponding to a cumulative probability of 97%. However, during rendering, the ellipsoids are scaled up by a factor of 10 in order to make them more easily visible. The major axis of the small ellipsoid in Figure 5 (left) is actually 0.32 mm; that of the larger ellipsoid is 1.84 mm.

The fixed sensor estimated the pose of the implant target ($^{\text{Optotrak}}_{\text{ImpTarg}}H$) with respect to the sensor, along with the corresponding covariance matrix. Using the known pose of the implant with respect to the implant target ($^{\text{ImpTarg}}_{\text{Implant}}H$), the pose of the implant with respect to the sensor was estimated, using

$$^{\text{Optotrak}}_{\text{Implant}}H = {}^{\text{Optotrak}}_{\text{ImpTarg}}H \; {}^{\text{ImpTarg}}_{\text{Implant}}H,$$

along with its covariance matrix.

Finally, the pose of the implant with respect to the HMD was estimated via

$$^{\text{Hmd}}_{\text{Implant}}H^{(\text{Opto})} = {}^{\text{Hmd}}_{\text{Optotrak}}H \; {}^{\text{Optotrak}}_{\text{Implant}}H.$$

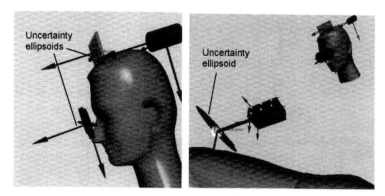

**Figure 5.** Uncertainties of poses derived from fixed sensor: (Left) HMD target (small ellipsoid) and HMD (large ellipsoid). (Right) Implant with respect to HMD.

The covariance matrix of this pose was estimated using Equation 9. The corresponding ellipsoid is shown in Figure 5 (right). The major axis of this ellipsoid is 8.23 mm. Note that the shape of this ellipsoid is elongated in the plane perpendicular to the line of sight, due to the orientation uncertainty in the HMD.

## Pose Estimation Using Head-Mounted Sensor

Using data from the head-mounted camera, we estimated the pose of the camera target ($^{\text{Camera}}_{\text{CamTarg}}H$) with respect to the camera, using the 2D-to-3D algorithm described earlier. From the estimated error in each 2D point measurement (0.5 pixel), the covariance matrix of the resulting pose was determined. Then, using the known pose of the implant with respect to the camera target ($^{\text{CamTarg}}_{\text{Implant}}H$), the pose of the implant with respect to the camera was estimated, via

$$^{\text{Camera}}_{\text{Implant}}H = {}^{\text{Camera}}_{\text{CamTarg}}H\, {}^{\text{CamTarg}}_{\text{Implant}}H.$$

The covariance matrix of the resulting pose was also estimated. The ellipsoids corresponding to the translational uncertainties are shown in Figure 6 (left). The major axis of the ellipsoid corresponding to $^{\text{Camera}}_{\text{CamTarg}}H$ is 24.6 mm. The major axis of the ellipsoid corresponding to the derived pose, $^{\text{Camera}}_{\text{Implant}}H$, is 19.9 mm.

Note the large uncertainty of $^{\text{Camera}}_{\text{CamTarg}}H$ along the line of sight to the camera, and very small uncertainty perpendicular to the line of sight. This is typical of poses that are estimated using the 2D-to-3D method. Intuitively, this may be explained as follows. A small translation of the object

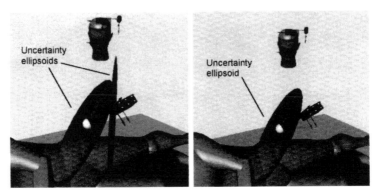

**Figure 6.** Uncertainties of poses derived from head camera: (Left) Camera target (long narrow ellipsoid) and implant with respect to camera (wide ellipsoid). (Right) Implant with respect to HMD.

parallel to the image plane results in an easily measurable change in the image, meaning that the uncertainty of translation is small in this plane. However, a small translation of the object perpendicular to the image plane generates only a very small image displacement, meaning that the uncertainty of translation is large in this direction.

Next, the pose of the implant with respect to the HMD is estimated, via

$$
{}^{\text{Hmd}}_{\text{Implant}}H^{(\text{cam})} = {}^{\text{Hmd}}_{\text{Camera}}H\, {}^{\text{Camera}}_{\text{Implant}}H.
$$

The covariance matrix of this pose was estimated using Equation 9. The corresponding ellipsoid is shown in Figure 6 (right). The major axis of this ellipsoid is 19.9 mm.

### Fusion of Data from Fixed and Head-Mounted Sensors

The two pose estimates, which were derived from the fixed and head-mounted sensors, can now be fused. Using Equation 10, we produce a combined estimate of the implant-to-HMD pose, along with the covariance matrix. The ellipsoids corresponding to the three poses, ${}^{\text{Hmd}}_{\text{Implant}}H^{(\text{opto})}$, ${}^{\text{Hmd}}_{\text{Implant}}H^{(\text{cam})}$, and ${}^{\text{Hmd}}_{\text{Implant}}H^{(\text{hybrid})}$ are shown in Figure 7. Note that the large ellipsoids, corresponding to ${}^{\text{Hmd}}_{\text{Implant}}H^{(\text{opto})}$ and ${}^{\text{Hmd}}_{\text{Implant}}H^{(\text{cam})}$, are nearly orthogonal. The ellipsoid ${}^{\text{Hmd}}_{\text{Implant}}H^{(\text{hybrid})}$, corresponding to the combined pose, is much smaller and is contained within the intersection volume of the larger ellipsoids. The right image of Figure 7 is a wire-frame rendering of the ellipsoids, which allows the smaller interior ellipsoid to be seen.

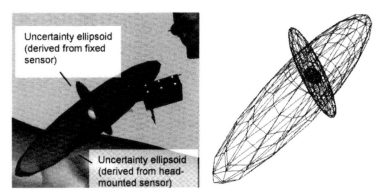

**Figure 7.** (Left) This figure depicts the fusion of the data. Note that the ellipsoids from the fixed sensor and the head-mounted sensor are nearly orthogonal. The ellipsoid corresponding to the resulting pose estimate is much smaller and is contained in the volume of intersection. (Right) This wire-frame rendering of the uncertainty ellipsoids allows the smaller (combined estimate) ellipsoid to be seen, contained in the intersection of the two larger ellipsoids.

The major axis corresponding to the uncertainty of the combined pose is only 1.47 mm.

## 5 Summary

This paper has developed a methodology for explicitly fusing sensor data from a combination of fixed and head-mounted sensors, in order to improve the registration of objects with respect to an HMD. The methodology was applied to an actual experimental augmented reality system. A typical configuration was analyzed, and it was shown that the hybrid system produces a pose estimate that is significantly more accurate than that produced by either sensor acting alone.

## Acknowledgments

This work was supported by a grant from Johnson & Johnson Professional, Inc.

The author would like to thank Dr. Tyrone Vincent for many helpful discussions, Khoi Nguyen for implementing many of the components of the experimental augmented reality system, and the anonymous reviewers for their helpful comments.

# References

[Azuma 94]      R. Azuma and G. Bishop. "Improving Static and Dynamic Registration in an Optical See-Through HMD." *Proceedings of 21st International SIGGRAPH Conference,* ACM, Orlando, FL, pp. 197–204, 1994.

[Azuma 97]      R.T. Azuma. "A Survey of Augmented Reality." *Presence,* Vol. 6, No. 4, pp. 355–385, 1997.

[Craig 90]      J. Craig. *Introduction to Robotics: Mechanics and Control,* 2nd ed. Addison-Wesley, Reading, MA, 1990.

[Haralick 93]   R. Haralick and L. Shapiro. *Computer and Robot Vision.* Addison-Wesley, Reading, MA, 1993.

[Hoff 96]       W.A. Hoff, T. Lyon, and K. Nguyen. "Computer Vision-Based Registration Techniques for Augmented Reality." *Proceedings of Intelligent Robots and Computer Vision XV,* Vol. 2904, SPIE, pp. 538–548, 1996.

[Horn 87]       B.K.P. Horn. "Closed-Form Solution of Absolute Orientation Using Unit Quaternions." *Journal of the Optical Society of America,* Vol. 4. No. 4, pp. 629–642, 1987.

[Kim 97]        D. Kim, S.W. Richards, and T.P. Caudell. "An Optical Tracker for Augmented Reality and Wearable Computers." *Proceedings of IEEE 1997 Annual International Symposium on Virtual Reality.* Albuquerque, NM, pp. 146–50, 1997.

[Trees 68]      H.L.V. Trees. *Detection, Estimation, and Modulation Theory.* New York: Wiley, 1968.

[Wang 90]       J.F. Wang, et al. "Tracking a Head-Mounted Display in a Room-Sized Environment with Head-Mounted Cameras." *Proceedings of Helmet-Mounted Displays II,* Vol. 1290, SPIE, pp. 47–57, 1990.

# Part IV

# The Hurdles of Computer Vision

*Despite its promise for AR applications, computer- vision-based registration must overcome a number of problems and limitations. The short papers and position statements in this section address various problems, including calibration, motion estimation, the use of natural features, and sensor fusion. The authors attempt to pave the way to possible solutions.*

# Augmenting Reality Without Camera or Scene Calibration

Jim Vallino

Early work in augmented reality grew out of virtual reality research domains. These initial augmented reality systems applied the same methods and technology as virtual reality systems to solve what at first appeared to be a similar problem: render correctly a scene of virtual objects as the user changes viewpoint in the world. This is indeed a shared goal of augmented and virtual reality systems. In virtual reality systems, our sense of presence is primarily controlled by the degree to which the visual stimulus presented corresponds to our kinesthetic senses. Augmented reality systems have an additional performance constraint: the correct registration between the user's view of the real scene and the virtual objects augmenting it. It is a more difficult task to maintain a compelling sense of presence when there are discrepancies between these two visual stimuli.

It is no mystery to researchers working in the area that the core problem in augmented reality is this accurate registration of the virtual computer-generated images with the user's view of the real scene. This is highlighted in most of the papers in the literature [Bajura 95] [Tuceryan 95] [State 96] [Holloway 97] [Azuma 94]. A taxonomy of registration methods used by augmented reality systems described in the literature is given in Table 1.

| Camera and/or Scene Calibration | Position Sensing | |
|---|---|---|
| | Yes | No |
| Yes | [Janin 93] [Feiner 93] [Rastogi 95] [Tuceryan 95] [State 96] | [Grimson 95] [Mellor 95] [Hoff 96] [Neumann 96] |
| No | [State 94] | [Uenohara 95] [Kutulakos 98] [Vallino 98] |

**Table 1.** Taxonomy of approaches to registration in augmented reality.

It is based on whether the system uses position sensing to monitor user location in the workspace and whether it requires calibration of the workspace and/or a video camera viewing the scene.

The accuracy of registration is measured by both static and dynamic error components. Static registration error provides a lower bound on accuracy. An augmented reality system will never perform better than its static accuracy. One contribution to static error is the method used for measurement of the user's viewpoint. In classic systems (if we may assume that the field is mature enough that some systems can be considered classics), viewpoint is measured by position sensing with magnetic sensors, such as the Polhemus sensor. Nonlinearities in the response of these sensors, due to the presence of large metal objects in the workspace, introduce a static registration error [Holloway 97] [Adelstein 92]. Systems using a video camera to view the scene require information about the camera's parameters, such as focal length, which must be obtained via careful calibration procedures. Any errors in calibration will be reflected in static errors also. The primary contributor to dynamic registration error is latency in the system [Holloway 97]. This latency comes not only from the time required to perform basic computational steps, but also latencies in position measurement and the time required to obtain the next frame of video in systems that incorporate cameras.

To achieve correct registration requires determination of the relationships among multiple coordinate systems [Tuceryan 95]: those of the world, the camera, and the virtual object. Using magnetic position sensors and metric calibration of cameras, these coordinate systems are referenced to a common world coordinate system, defined as a standard Euclidean coordinate system. This refernce system has the advantage of being easy to conceptualize, and one can take out a tape measure and measure locations in the world. The downside is that you require this metric information in order to compute the necessary relationships between coordinate systems. A desire to simplify the registration process by eliminating the need for this metric information along with all position measurement and camera calibration is what motivated much of my thesis work in augmented reality at the University of Rochester [Vallino 98]. Augmented reality was a natural application of recent work in computer vision research that extracted structure and motion from scenes using uncalibrated cameras [Koenderink 91] [Mundy 92] [Weinshall 93].

The technique of augmenting reality using affine representations relates all the coordinate systems to a common non-Euclidean affine coordinate system. The definition of this coordinate system is obtained at runtime from the projections of four non-coplanar feature points that are tracked through the video sequence. One of the absolute beauties of this approach is the simplicity of the mathematics. An augmented reality system ultimately needs to compute the projection matrix that the computer graphics camera uses to render the virtual objects. Using affine representations, the projection matrix is created directly from the projection of feature points in a video image. Even after working steadily in this environment for several years, I am still amazed that it works as well as it does. At this point, one must ask the question: "Can this metric-free method scale up and be viable for the long term such that it warrants continued work?" My answer would be a qualified yes.

This method definitely has its disadvantages. One can argue that the simplicity of computing the projection matrix is offset by the additional requirement of tracking feature points in real time. (Note that any augmented reality technique based on computer vision methods [State 96] [Mellor 95] [Hoff 96] [Neumann 96] has this additional requirement.) Tracking of arbitrary targets is still an open problem in computer vision research. If, however, you are willing to engineer your problem with regard to the particular features being tracked, the technology is available for accurate real-time tracking of feature points. In my own work, I engineered my experiments to minimize the tracking problem by using color segmentation to track features. Moving to a more natural setting will require additional work on the feature tracking subsystem.

This non-Euclidean method defines coordinates as the linear combination of projections of feature points. Since this coordinate system is only computed at runtime, a priori placement of virtual objects cannot be performed. Applications operating in an unknown environment will, in general, be required to do runtime placement of virtual objects and may be particularly well suited to this method.

The University of Rochester is investigating one such application. It is a military or crime interdiction video surveillance and monitoring scenario, depicted in Figure 1. Here the system generates an augmented view of a scene in a command and control center, based on ground and aerial surveillance and detection of common feature points. One can also envision another scenario in which aerial surveillance is used to augment a ground-fighter's view of a battle scene.

Approximating the true perspective operation of a video camera with an affine camera [Mundy 92] introduces errors, particularly when the distance from camera to object diminishes. This will cause the system to produce a static registration error. To mitigate this error, the use of projective representations should be researched [Faugeras 92]. Affine representations have their place and have been found to provide more accurate reprojection

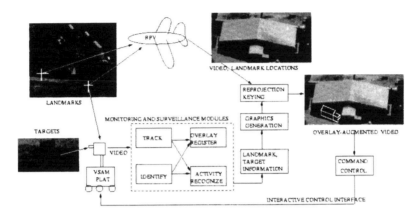

**Figure 1.** Video surveillance and monitoring application.

results for large-object-to-camera distances [Wiles 96]. An adaptive technique that automatically switches to the most appropriate representation, as the camera moves or its lens is zoomed, may yield the best results.

It should come as no surprise that an augmented reality system based on affine representations is not immune to latency problems generating dynamic errors. In my system, I measured system latencies on the order of 70 to 90 ms. or two to three frames of video. Without metric data or position sensing, many of the approaches [Zikan 94] for minimizing errors due to latency cannot be directly applied.

In my testing, I found that simple filtering and forward prediction applied directly to the projection of the feature points in the image yielded marked improvement in registration. These results are shown in Figure 2. The mean Euclidean pixel error between a physical point moving in the scene and its reprojected virtual point is shown on the $y$-axis. The $x$-axis is the number of frames of forward prediction applied to the feature point locations, assuming that feature motion is modeled by constant velocity in the image. The several plots are for different filtering methods applied to the feature points for noise reduction. Across-the-board improvements were seen for two- and three-frame forward prediction of feature point projections. These results show that methods are available for improving latency in a system using non-metric representations. On the graphics side, there are limitations with this method due to the nature of the projection matrices computed. In general, the affine projection matrix will not be orthonormal. For many standard computer graphic techniques, such as lighting computations, an orthonormal system is a requirement. If more

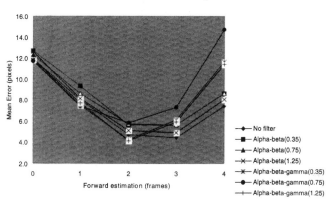

**Figure 2.** Dynamic registration error reduction

photorealistic rendering of the virtual objects is required, then additional research must be undertaken to determine the correct method for executing these computer graphics algorithms in a common affine coordinate system.

The use of affine representations is one end of the spectrum of computer-vision-based methods for implementing augmented reality systems. Despite its disadvantages, it is an attractive method that has potential for certain applications. Applications that will work in an unknown environment where a priori calibration and measurement will not be possible seem particularly well suited to the method. There are potential avenues for research that might improve the method's performance. So far, there has been little work [State 96] that tries to integrate the different methods for performing registration in order to optimize the overall result. Further research in that area might yield synergies that are currently unknown.

# References

[Adelstein 92]   B.D. Adelstein, E.R. Johnston, and S.R. Ellis. "A Testbed for Characterizing Dynamic Response of Virtual Environment Spatial Sensors." *Proceedings of 5th Annual Symposium on User Interface Software and Technology,* Monterey, pp. 15–21, 1992.

[Azuma 94]   R. Azuma and G. Bishop, "Improving Static and Dynamic Registration in an Optical See-through HMD," *Proceedings of SIGGRAPH '94,* pp. 197–204, 1994.

[Bajura 95]     M. Bajura and U. Neumann. "Dynamic Registration Correction in Video-Based Augmented Reality Systems." *IEEE Computer Graphics and Applications,* Vol. 15, pp. 52–60, 1995.

[Faugeras 92]   O.D. Faugeras. "What Can be Seen in Three Dimensions with an Uncalibrated Stereo Rig?" *Proceedings of Second European Conference on Computer Vision,* pp. 563–578, 1992.

[Feiner 93]     S. Feiner, B. MacIntyre, and D. Seligmann. "Knowledge-Based Augmented Reality." *Communications of the ACM,* Vol. 36, pp. 53–62, 1993.

[Grimson 95]    W.E.L. Grimson, G.J. Ettinger, S.J. White, P.L. Gleason, T. Lozano-Perez, M.W. Wells III, and R. Kikinis. "Evaluating and Validating an Automated Registration System for Enhanced Reality Visualization in Surgery." *Proceedings of Computer Vision, Virtual Reality, and Robotics in Medicine '95,* Nice, France, pp. 3–12, 1995.

[Hoff 96]       W.A. Hoff, K. Nguyen, and T. Lyon, "Computer Vision-Based Registration Techniques for Augmented Reality." *Proceedings of SPIE Vol. 2904: Intelligent Robots and Computer Vision XV: Algorithms, Techniques, Active Vision, and Materials Handling,* Boston, MA, pp. 538–548, 1996.

[Holloway 97]   R.L. Holloway. "Registration Error Analysis for Augmented Reality." *Presence,* Vol. 6, pp. 413–432, 1997.

[Janin 93]      A.L. Janin, D.W. Mizell, and T.P. Caudell. "Calibration of Head-mounted Displays for Augmented Reality Applications." *Proceedings of IEEE Virtual Reality Annual International Symposium '93,* Seattle, WA, pp. 246–255, 1993.

[Koenderink 91] J.J. Koenderink and A.J. van Doorn. "Affine Structure from Motion." *Journal of the Optical Society of America A,* Vol. 8, pp. 377–385, 1991.

[Kutulakos 98]  K.N. Kutulakos and J.R. Vallino. "Calibration-Free Augmented Reality." *IEEE Transactions on Visualization and Computer Graphics,* Vol. 4, pp. 1–20, 1998.

[Mellor 95]     J.P. Mellor. "Enhanced Reality Visualization in a Surgical Environment." Masters Thesis, AI Lab, Massachusetts Institute of Technology, Cambridge, MA, 1995.

[Mundy 92]      J.L. Mundy and A. Zisserman. *Geometric Invariance in Computer Vision.* Cambridge, MA: The MIT Press, 1992.

[Neumann 96] U. Neumann and Y. Cho. "A Self-Tracking Augmented Reality System," *Proceedings of ACM Symposium on Virtual Reality Software and Technology,* pp. 109–115, 1996.

[Rastogi 95] A. Rastogi, P. Milgram, and J.J. Grodski. "Augmented Telerobotic Control: A Visual Interface for Unstructured Environments," http://vered.rose.utoronto.ca/people/anu_dir/papers/atc/atcDND.html, 1995.

[State 94] A. State, D.T. Chen, C. Tector, A. Brandt, H. Chen, R. Ohbuchi, M. Bajura, and H. Fuchs. "Case Study: Observing a Volume Rendered Fetus within a Pregnant Patient." *Proceedings of the 1994 IEEE Visualization Conference,* Washington, D.C., pp. 364–368, 1994.

[State 96] A. State, G. Hirota, D.T. Chen, W.F. Garrett, and M.A. Livingston. "Superior Augmented Reality Registration by Integrating Landmark Tracking and Magnetic Tracking," *Proceedings of the ACM SIGGRAPH Conference on Computer Graphics,* New Orleans, pp. 429–438, 1996.

[Tuceryan 95] M. Tuceryan, D.S. Greer, R.T. Whitaker, D.E. Breen, C. Crampton, E. Rose, and K.H. Ahlers. "Calibration Requirements and Procedures for a Monitor-Based Augmented Reality System," *IEEE Transactions on Visualization and Computer Graphics,* Vol. 1, pp. 255–273, 1995.

[Uenohara 95] M. Uenohara and T. Kanade. "Vision-Based Object Registration for Real-Time Image Overlay." *Computer Vision, Virtual Reality and Robotics in Medicine: CVRMed '95, Lecture Notes in Computer Science,* edited by N. Ayache, Berlin: Springer-Verlag, pp. 14–22, 1995.

[Vallino 98] J.R. Vallino. "Interactive Augmented Reality," PhD Thesis, Department of Computer Science, University of Rochester, 1998.

[Weinshall 93] D. Weinshall and C. Tomasi. "Linear and Incremental Acquisition of Invariant Shape Models from Image Sequences." *Proceedings 4th IEEE International Conference on Computer Vision,* pp. 675–682, 1993.

[Wiles 96] C. Wiles and M. Brady. "On the Appropriateness of Camera Models." *Proceedings of the Fourth European Conference on Computer Vision,* Cambridge, UK, pp. 228–237, 1996.

[Zikan 94] K. Zikan, W.D. Curtis, H.A. Sowizral, and A.L. Janin. "Note on Dynamics of Human Head Motions and on Predictive Filtering of Head-set Orientations." *Proceedings of SPIE Vol. 2351: Telemanipulator and Telepresence Technologies,* pp. 328–336, 1994.

# Visual Servoing-based Augmented Reality

Venkataraman Sundareswaran and Reinhold Behringer

## 1  Introduction

Augmented reality (AR) mixes computer-generated, synthetic elements (3D/2D graphics, 3D audio) with the real world in such a way that synthetic elements appear to be part of the real world. Various techniques are used to accomplish this, including magnetic tracking of position and orientation, and video-based tracking. This paper focuses on video-based AR, in which a camera is used to generate an image of the real world; the image is then processed to determine where the computer-generated elements should be displayed. Video-based AR is particularly popular among AR researchers because of the accuracy that can be achieved in video image processing.

The central problem in video-based AR is registration — the alignment of 3D graphical information with a real scene. If the relative position and orientation of the camera is known, the graphical rendering can be done correctly.

## 2  Background

There are three major approaches to solving the registration problem: (1) object-pose estimation methods; (2) observer-pose estimation methods; and (3) camera-motion estimation methods.

Object-pose estimation methods determine the position and orientation of the object (e.g., a plane containing landmarks such as LED patterns (MIT) or colored dots [Neumann 96]). These methods are based on pose determination schemes such as that of Fischler and Bolles [Fischler 81]. Typically, the landmarks are located using image processing, and the pose is determined in each frame. The pose is then used to render the 3D model. Registered templates [Uenohara 95] and affine coordinate system-based approaches [Kutulakos 96] are also used to determine object pose.

Magnetic tracking can readily provide the position and orientation of the observer (e.g., [Webster 96] [State 96].) The major limitations of magnetic tracking are its short range (typically an eight-ft. radius) and sensitivity to metallic objects in the vicinity. Video-based observer-pose estimation methods attempt to compute the position and orientation of the camera from the position of landmarks in the images. In a surgical application, Grimson et al. [Grimson 96] used a data-model minimization, a least-squares minimization of distance between the image data and 3D model data obtained *a priori* by scanning with a laser rangefinder. Using the "Hung-Yeh-Harwood pose estimation method," Hoff et al. [Hoff 96], at the Colorado School of Mines, developed an observer-pose estimation from concentric circle markers. Work at UNC [State 96] integrated magnetic tracking and video-based observer-pose estimation in order to demonstrate a robust system for AR.

Our approach falls into the third category: motion estimation. The general problem of reliable 3D motion estimation from image features is a largely unsolved computer-vision problem. However, by restricting it to the sub-problem of easily identifiable landmarks, we can solve the motion estimation problem. Koller et al. [Koller 97] used a linear acceleration model for camera motion to determine the motion of the camera. Our approach is based on the mathematical formalism of visual servoing, which will be explained in the next section.

## 3   3D Tracking

In this section, we describe an algorithm for tracking an object in three dimensions based on 2D coordinates of object features and a known 3D model of the object. The algorithm is based on principles from visual servoing.

### 3.1   Visual Servoing

Visual servoing is controlling a system based on processed visual information — typically, a robot end-effector. It is a well-developed theory for robotic vision [Espiau 92] [Feddema 89] [Papanikolopoulos 93] [Sundareswaran 96] [Weiss 87]. Visual servoing is carried out in a closed-loop fashion, as shown in Figure 1.

We would like the set of system states $s$ to attain certain target values $s_r$. The current values of the states $s$ are measured by a camera viewing the scene. The system uses the error (difference between the target values and current values) to determine the motion parameters $T$ and $\Omega$, moving the camera in order to reduce error. We adopt the standard coordinate systems shown in Figure 1. The translational velocity $T$ has components $U$, $V$, and $W$. The components of the rotational velocity $\Omega$ are $A$, $B$,

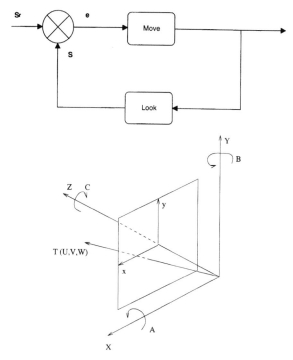

**Figure 1.** Schematic of the visual servoing approach, and the coordinate system.

and $C$. We need to know the analytical relationship between the motion parameters and the state **s**. Usually, the forward relationship, namely the change in **s** due to parameters $T$ and $\Omega$, is known. The goal is to minimize $\|s - s_r\|$. Let us define the error function

$$\mathbf{e} = \mathbf{s} - \mathbf{s}_r; \tag{1}$$

the change in error is given by

$$\dot{\mathbf{e}} = \dot{S}_r.$$

We would like the error function to decay exponentially:

$$\dot{\mathbf{e}} = -\lambda \cdot \mathbf{e},$$

where $\lambda$, the constant in the exponential, controls the decay rate (i.e., speed of convergence); therefore, $\dot{\mathbf{s}} = -\lambda \cdot (\mathbf{s} - \mathbf{s}_r)$. From standard optic flow equations (e.g., [Horn 87]), we know that we can write the 2D displacement of an image feature at $(x_p, y_p)$ as

$$
\begin{aligned}
\dot{x}_p &= \tfrac{1}{Z(x_p, y_p)} \left[ -U + x_p W \right] + A x_p y_p - B \left[ 1 + x_p^2 \right] + C y_p, \\
\dot{y}_p &= \tfrac{1}{Z(x_p, y_p)} \left[ -V + y_p W \right] + A \left[ 1 + y_p^2 \right] - B x_p y_p - C x_p.
\end{aligned}
\tag{2}
$$

We assume that the images are planar, obtained by pin-hole perspective approximation with a focal length of one (see Figure 1). The relationship between the change in the 2D projection of a point and the motion parameters is

$$
\dot{s} = L \begin{pmatrix} T \\ \Omega \end{pmatrix},
\tag{3}
$$

where $L$ is the "interaction matrix," consisting of 2D coordinates $(x_p, y_p)$ and the depth $Z$ of the 3D point projected at $(x_p, y_p)$; $T$ is the translation vector; and $\Omega$ is the rotation vector. We would like to determine $T$ and $\Omega$. Assuming that the motion of features $s$ is due to the motion $T$ and $\Omega$, we obtain

$$
L \begin{pmatrix} T \\ \Omega \end{pmatrix} = -\lambda \mathbf{e}.
\tag{4}
$$

Inverting Equation 4, we get the control law

$$
\begin{pmatrix} T \\ \Omega \end{pmatrix} = -\lambda L^+ \mathbf{e},
\tag{5}
$$

where $L^+$ is the pseudo-inverse of $L$.

This allows us to compute the motion of the camera required to minimize the error $e$. When performed in closed-loop, the value $s$ will reach $s_r$ when error $e$ is reduced to zero.

## 3.2   Application to AR

The goal of the AR process is to maintain the alignment between image landmarks and the graphical rendering of these landmarks. In other words, we would like to minimize the error between these two sets of values, yielding the following choices: the target values $(s_r)$ are measured from the image using image processing; the current values $(s)$ are known from the graphics rendering process; and the control (computed from Equation 5) is applied to the virtual camera that renders the 3D scene. The goal of the control process is to minimize the difference between the image location of the landmarks and the rendered position of the landmarks. Obviously, if successful, this control process will achieve the registration of the image and the model. If operated continuously, the control process will keep the model registered with the image as the camera is moved about.

## 4 Experimental Results

We have implemented the visual servoing-based AR on a PC with a 200-MHz Pentium Pro processor, an Imaging Technology framegrabber, and an OpenGL accelerator card. The landmarks are circular concentric ring markers. Each marker has a unique internal structure, and this identifies the marker uniquely in images. The circular shape simplifies the image processing involved in locating and identifying the markers. The overall flow diagram of the system is shown in Figure 2.

The image is grabbed by the Imaging Technology framegrabber, and the processing is carried out in the CPU. The graphical rendering is done using World Tool Kit (WTK). The system runs at 8–10 fps. The registration of the 3D wireframe model of a PC is shown in Figure 2.

To initialize the system, we choose several candidate viewpoints around the object (six, for the example shown in Figure 2), and compute the minimization error for each candidate. We choose the candidate viewpoint with the least error and converge from there. For complicated objects, we will have to choose numerous candidate viewpoints to ensure that the initial guess is close to the actual viewpoint. This is not a problem, since it only needs to be done once, at start-up. After that, during tracking, the pose from the previous frame is used as the initial guess.

The $\lambda$ value is chosen in an adaptive fashion based on the error $e$. When error $e$ is large, we use a relatively small $\lambda$ value, and increase it as $e$ decreases. This allows for stable convergence. The range of values used for $\lambda$ is from 0.25 to 0.9. Large error values will be present at start-up and when there is a loss of tracking (due to visual obstruction or a sudden jerk of the camera); at other times, a high $\lambda$ value will be used, ensuring fast operation.

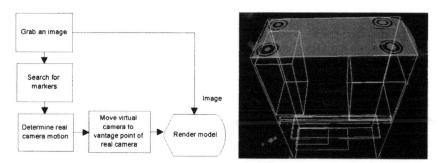

**Figure 2.** Flow diagram of the system, and a sample frame (full sequence at http://hci.rsc.rockwell.com/3Dreg.shtml)

# 5   Discussion

Minimizing the distance between image landmarks and their counterparts in the model projection is, by definition, the registration requirement. This minimization is carried out directly by the approach described in this paper. Also, since loss of alignment occurs due to motion, computing motion in order to reduce misalignment is a direct solution. The six motion parameters are computed, and the virtual camera is "moved" within these parameters. Since the computation is carried out in a closed loop, the motion of the camera is imitated by the virtual camera rendering the graphics.

The approach is independent of the type of landmark, as long as at least four of them can be detected and identified. Three landmarks are required in order to determine the six motion parameters (three landmarks yield two equations each, as in Equation 2). However, to resolve the ambiguity of planar orientation, we need at least four landmarks. When more than four landmarks can be detected, we have an over-determined system and the redundancy increases robustness.

In the description of Equation 3, we note that the computation requires the $Z$-values of the landmarks. These values are obtained from the rendered scene, since the geometry of the model and the position of the viewpoint are known. This approximation is valid if the tracking is fast and accurate enough.

Our implementation runs at 8-10 fps, but this speed can be improved by: replacing the World-Tool-Kit-based rendering by a less "bulk" library; performing some or all of the image processing in hardware; and using a faster CPU (we currently use a 200-MHz Pentium).

# 6   Conclusions

In this paper, we presented a new approach to solving the registration problem in vision-based AR — using the visual servoing technique. We have implemented this technique on desktop systems, and a simpler version on wearable and handheld devices.

# References

[Espiau 92]                    B. Espiau, F. Chaumette, and P. Rives. "A New Approach to Visual Servoing in Robotics." *IEEE Transactions on Robotics and Automation*, Vol. 8, No. 3, pp. 313–326, 1992.

[Feddema 89]           J.T. Feddema and O.R. Mitchell. "Vision-guided Servo-
                       ing with Feature-based Trajectory Generation." *IEEE
                       Transactions on Robotics and Automation*, Vol. 5, No.
                       5, pp. 691–700, 1989.

[Fischler 81]          M.A. Fischler and R.C. Bolles. "Random Sample Con-
                       sensus: A Paradigm for Model Fitting with Applica-
                       tions to Image Analysis and Automated Cartography."
                       *Graphics and Image Processing*, Vol. 24, No. 6, pp. 381–
                       395, 1981.

[Grimson 96]           W.E.L. Grimson, G.J. Ettinger, S.J. White, T, Lozano-
                       Perez, W.M. Wells III, and R. Kikinis. "An Automatic
                       Rregistration Method for Frameless Stereotaxy, Image
                       Guided Surgery, and Enhanced Reality Visualization."
                       *IEEE Transactions on Medical Imaging*, Vol. 15, No. 2,
                       pp. 129–140, 1996.

[Hoff 96]              W.A. Hoff, T. Lyon, and K. Nguyen. "Computer Vision-
                       based Registration Techniques for Augmented Reality."
                       *Proceedings of Intelligent Robotics and Computer Vision
                       XV*, Vol. 2904 in *Intelligent Systems and Advanced Man-
                       ufacturing, SPIE*, Boston, Massachusetts, pp. 538–548,
                       1996.

[Horn 87]              B.K.P. Horn. *Robot Vision*. Cambridge:   The MIT
                       Press, 1987.

[Koller 97]            D. Koller, G. Klinker, E. Rose, D. Breen, R. Whitaker,
                       and M. Tuceryan. "Real-time Vision-Based Camera
                       Tracking for Augmented Reality Applications." *Proceed-
                       ings of the ACM Symposium on Virtual Reality Software
                       and Technology (VRST-97)*, Lausanne, Switzerland, pp.
                       87–94, 1997.

[Kutulakos 96]         K.N. Kutulakos and J. Vallino. "Affine Object Repre-
                       sentations for Calibration-free Augmented Reality." *Pro-
                       ceedings of IEEE Virtual Reality Annual Symposium
                       (VRAIS)*, Santa Clara, CA, pp. 25–36, 1996.

[Neumann 96]           U. Neumann and Y. Cho. "A Self-Tracking Augmented
                       Reality System." *Proceedings of the ACM Symposium
                       on Virtual Reality Software and Technology*, Hong Kong,
                       pp. 109–115, 1996.

[Papanikolopoulos 93]   N. Papanikolopoulos, P. Khosla, and T. Kanade. "Visual Tracking of a Moving Target by a Camera Mounted on a Robot: A Combination of Control and Vision." *IEEE Transactions on Robotics and Automation*, Vol. 9, No. 1, pp. 14–35, 1993.

[State 96]   A. State, G. Hirota, D.T. Chen, W.F. Garrett, and M.A. Livingston. "Superior Augmented-Reality Registration by Integrating Landmark Tracking and Magnetic Tracking." *Proceedings of SIGGRAPH '96*, New Orleans, pp. 429–438, 1996.

[Sundareswaran 96]   V. Sundareswaran, P, Bouthemy, and F. Chaumette. "Exploiting Image Motion for Active Vision in a Visual Servoing Framework." *International Journal of Robotics Research*, Vol. 15, No. 6, pp. 629–645, 1996.

[Uenohara 95]   M. Uenohara and T. Kanade. "Vision-based Object Registration for Real-time Image Overlay." *International Journal of Computers in Biology and Medicine,* Vol. 25, No. 2, pp. 249–260, March 1995.

[Webster 96]   A. Webster, S. Feiner, B. MacIntyre, W. Massie, and T. Krueger. "Augmented Reality in Architectural Construction, Inspection, and Renovation." *Proceedings of the Third ASCE Congress for Computing in Civil Engineering*, Anaheim, CA, pp. 913–919, June 1996.

[Weiss 87]   L.E. Weiss, A.C. Sanderson, and C.P. Neuman. "Dynamic Sensor-based Control of Robots with Visual Feedback." *IEEE Transactions on Robotics and Automation*, Vol. 3, No. 5, pp. 404–417, 1987.

# Constrained Self-Calibration for Augmented Reality Registration

Jeffrey Mendelsohn, Kostas Daniilidis, and Ruzena Bajcsy

**Abstract.** *A novel technique is presented for estimating camera calibration and trajectory from a set of targets arbitrarily placed in the environment. This algorithm can be used to replace other registration systems requiring a priori camera calibration or the use of unwieldy magnetic trackers. Experimental results validate the accuracy of registration in dynamic video sequences.*

## 1 Introduction

A critical issue in augmented reality is the *registration* problem, which refers to determining the proper alignment of the real and virtual worlds. The importance of this task is illustrated in [Azuma 97] and [Mellor 95], which describe an application in which a virtual tumor is to appear on a healthy organ during surgery. Registration errors in video-based systems are classified, according to [Azuma 97], into *static* and *dynamic* errors. Static errors arise from pose estimation, intrinsic parameter perturbations, and optical distortions. Dynamic registration discrepancies are caused by the delay due to video acquisition, vision computations, and graphical rendering.

We address here the so-called static registration errors. Earlier systems used a priori calibration of the camera parameters and recovered the observer's pose with a magnetic tracker or infrared tracker with landmarks on the observer. Such systems are restricted to close-range environments without magnetic interferences. Vision techniques can alleviate these weaknesses and offer a kind of direct feedback; the image data used for visualization and merging are also used for solving the registration problem. This fact was recognized early by Bajura [Bajura 95] and Mellor [Mellor 95]. Both of their approaches, as well as the more sophisticated one proposed by Koller et al. [Koller 97], use off-line calibration of intrinsic camera parameters and reference landmarks which are accurately distributed in space. Although these algorithms can perform very accurately, they suffer from practical restrictions: they cannot improve the initial estimation of camera parameters and require a distribution of fiducial targets with known relative positions. In long-range, outdoor terrain applications, such a landmark configuration is difficult to achieve.

Other techniques include structure-from-motion (for a review see [Huang 94]), self-calibration [Maybank 92], and calibration (see [Tsai 89] for a review) followed by pose estimations [Liu 90]. Use of the structure-from-motion technique with known calibration produces a reconstruction up to a similitude transformation, and using it with unknown calibration produces a reconstruction up to a projective transformation; this limitation may not be acceptable. Relying on calibration and pose estimation implies the use of a target which must be visible over the entire workspace. This alone tends to make it infeasible for AR. A further complication is degeneration to the recovery of only an affine transformation if the target within the image is small [Kumar 96]. Self-calibration approaches are known to suffer under the usual problems of structure-from-motion, as well as the confusion between intrinsic parameters and motion [Oliensis 98].

Here we propose a technical solution to decrease the sensitivity of self-calibration by placing small, easily identifiable targets of known shape in the environment. Assuming an appropriate ratio of size to distance, these targets fix the scale and resolve known ambiguities. The proposed algorithm can be applied in any moving camera application, and enables the direct merging of several views into a single 3D-representation of the type required during a stereo reconstruction of a room. For other applications, the path of the observer is recovered and can be used for global navigation.

## 2   Registration Problem

First the relationship between a world point $x_w$ and its projection in frame $f$ must be made explicit. Denote the rigid displacement from the $f$th frame to the world coordinate system by rotation $R_f$ and translation $t_f$: $x_f = R_f x_w + t_f$. A projection and an affine transformation produce the image point $p_f$ of frame point $x_f$:

$$p_f = \begin{bmatrix} fs & 0 & c_u \\ 0 & f & c_v \\ 0 & 0 & 1 \end{bmatrix} \frac{x_f}{\hat{z}^T x_f} = C \frac{x_f}{\hat{z}^T x_f} \quad .$$

The intrinsic camera parameters are the focal length $f$, image center coordinates $c_u$ and $c_v$, and a scale factor $s$ accounting for pixel aspect ratio and sampling rate discrepancies. Combining the mappings:

$$\mathbf{p_f} = \mathbf{C}\frac{\mathbf{R_f x_w} + \mathbf{t_f}}{\hat{\mathbf{z}}^T(\mathbf{R_f x_w} + \mathbf{t_f})} \quad . \tag{6}$$

The goal of registration, as presented in this equation, is the estimation of $\mathbf{C}$ and all $\mathbf{R_f}$ and $\mathbf{t_f}$ from the observations $\mathbf{p_f}$.

Here we propose an approach which we call constrained self-calibration; this approach uses minimal reference information in order to reduce greatly the degrees of freedom. Specifically, laser-printed targets will be placed in the environment with unknown relative displacements. A target constrains the solution space by replacing a set of unknown point coordinates with an unknown rigid transformation of known points: $\mathbf{x_w} = \mathbf{R_t x_t} + \mathbf{t_t}$. For instance, a target of five points has only six degrees of freedom (the rotation and translation) instead of 15 unknown world coordinates. Another advantage to using targets is that we can incorporate a numbering system to identify each target uniquely and hence solve the correspondence problem (matching a target in one frame to the same target in another frame). Substituting for $\mathbf{x_w}$ in (6):

$$\mathbf{p_{ft}} = \mathbf{C}\frac{\mathbf{R_f}(\mathbf{R_t x_t} + \mathbf{t_t}) + \mathbf{t_f}}{\hat{\mathbf{z}}^T[\mathbf{R_f}(\mathbf{R_t x_t} + \mathbf{t_t}) + \mathbf{t_f}]} \quad . \tag{7}$$

Now the goal of registration is the estimation of the intrinsics $\mathbf{C}$, all rotations $\mathbf{R_f}$, and all translations $\mathbf{t_f}$ from the observations $\mathbf{p_{ft}}$ and the known $\mathbf{x_t}$.

## 3 Algorithm Description

The task will be subdivided into three steps: the localization of the targets in an image as well as the establishment of target correspondences across images; the extraction of the projected data; and the optimization of the desired parameters with regard to the data.

The targets we used were designed to be easy to find in an image and were numbered for easy identification. A target is essentially a laser-printed, thick-walled, black pentagon with a white center. The requirements of a white center and five sides provided verification that each dark blot was truly a target. To ensure proper estimation of the orientation of the target, one of the sides of the pentagon was colored dark gray instead of black; this created a bottom to the target. Given the projection of the five corners, a transformation from target coordinates to image coordinates was computed in order to provide a method for localizing the numbering system.

From the projection transformation, the pixels corresponding to the step-edge between the black pentagon and the white paper were computed. These were adjusted by a one-dimensional search — roughly perpendicular to the orientation of the line — for the actual edge coordinates. The edge coordinates became data for estimating the slope and intercept of each line as well as the covariances of these values. These line values were used instead of corner points for greater accuracy during the optimization.

Given the line data from each target in each frame and the target labeling, the calibration and transformation parameters were found using the Levenberg-Marquardt algorithm in order to minimize a least-squares error metric based on the line data and the projection model for lines. The model is derivable in two steps. First, we derive the projection of a point through all the transformations; and second, the projection of a target line is derived from the point projection equation.

A target point is first transformed into the first frame's coordinate system and then into the current frame's system:

$$\mathbf{x_f} = \mathbf{R_f}\left(\mathbf{R_t}\mathbf{x_t} + \mathbf{t_t}\right) + \mathbf{t_f}$$

For notational simplicity, define $\mathbf{R} \equiv \mathbf{R_f}\mathbf{R_t}$ and $\mathbf{t} \equiv \mathbf{R_f}\mathbf{t_t} + \mathbf{t_f}$. In the case of a stereo configuration, the equations for the additional camera are obtained if $\mathbf{R} \equiv \mathbf{R_s}\mathbf{R_f}\mathbf{R_t}$ and $\mathbf{t} \equiv \mathbf{R_s}(\mathbf{R_f}\mathbf{t_t} + \mathbf{t_f}) + \mathbf{t_s}$, where the relative orientation between the two cameras is given by $\mathbf{R_s}$ and $\mathbf{t_s}$. Given that the target is planar, its projection equations can be written as:

$$\mathbf{p_{ft}} = \begin{bmatrix} u \\ v \\ 1 \end{bmatrix} = \mathbf{CHx_t}\frac{1}{\hat{\mathbf{z}}^T\mathbf{Hx_t}}, \qquad \text{where} \qquad \mathbf{H} \equiv \begin{bmatrix} r_{11} & r_{12} & t_x \\ r_{21} & r_{22} & t_y \\ r_{31} & r_{32} & t_z \end{bmatrix}$$

$\mathbf{H}$ is the homography between the image and the target plane.

Lines in the target were defined by a point $\mathbf{x_l}$ and a direction vector $\mathbf{d}$. A point $\mathbf{x_t}$ of this line is then parameterized by a distance $\alpha$ from $\mathbf{x_l}$:

$$\mathbf{x_t} = \mathbf{x_l} + \alpha\hat{\mathbf{d}}$$

After the projection:

$$u = fs\frac{\mathbf{h_1^T}\left(\mathbf{x_l} + \hat{\mathbf{d}}\alpha\right)}{\mathbf{h_3^T}\left(\mathbf{x_l} + \hat{\mathbf{d}}\alpha\right)} + c_u \qquad\qquad v = f\frac{\mathbf{h_2^T}\left(\mathbf{x_l} + \hat{\mathbf{d}}\alpha\right)}{\mathbf{h_3^T}\left(\mathbf{x_l} + \hat{\mathbf{d}}\alpha\right)} + c_v \qquad (8)$$

$$\text{where:} \qquad \mathbf{H^T} \equiv \begin{bmatrix} \mathbf{h_1} & \mathbf{h_2} & \mathbf{h_3} \end{bmatrix}$$

Using one of the equations to solve for $\alpha$ and substituting into the other equation, we may obtain a single constraint. Which equation to substitute into should be determined by the general orientation of the image line so as to ensure that the in-image line estimation is a well-posed problem. The resultant constraint is squared to provide a least-squares error metric for optimally estimating the line parameters assuming additive Gaussian noise.

Of primary importance is the maintenance of the orthonormality of the rotation matrices. The algorithm implements the Rodriguez representation in order to ensure that the degrees of freedom for rotation are only three.

## 4 Experiments

A 17-frame image sequence was taken at the GRASP lab to demonstrate the algorithm. The visual system consisted of two cameras rigidly mounted to a horizontal link that moved through the scene so as to maintain a table in the right camera's field of view. The scene consisted primarily of a table and two walls on which planar targets — the pentagonal objects in Figure 1 — were hung.

In total, eight targets were imaged by the cameras. Seven of these were visible in the first stereo pair, and, thereafter, typically three or four of the targets were clearly visible. The displacement of the targets from the corner ranged up to approximately 2.25 meters on the left wall and 2 meters on the right. The targets are 19.5 centimeters tall.

The first results are a set of augmented reality images shown in Figure 1. To create these, the dimensions of the table were measured relative to a coordinate system defined by the two targets 'on' the table. The line through

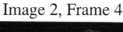

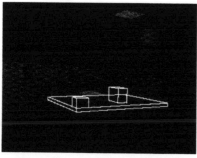 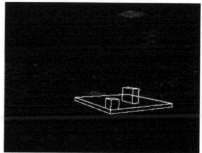

**Figure 1.** Both results are projections into image number two of the left camera. The left is the result after four, and the right after 16, frames of processing.

**Figure 2.** Bird's eye view of reconstruction. Targets are represented with Ts and the observer with a schematic. Past observer positions show trajectory.

both targets' origins was chosen as the $x$-axis and the average of their normals as the $z$-axis. The system's origin coincides with the left target's origin. Given these relative measurements, the reconstruction provides a mapping into the coordinate system of each camera and hence allows for projection. Two boxes were also projected into the images. The initially huge errors present in the projection are due to the implausibility of separating calibration and pose values from the small number of seen targets and the positions they occupy; although they are not presented, the reprojection errors for the targets is *smallest* during these frames, since the ratio of degrees of freedom to observed data is highest.

Another qualitative result is presented in the bird's eye view of the initial and final reconstructions in Figure 2. The targets were hung in a corner, so the true configuration contains two lines that are perpendicular, as they appear in Frame 17. The improvement in the reconstruction is clearly evident in the angle created by the two sets of targets.

Quantitatively, two results are provided. The first is a comparison of the estimated separation between the cameras of the stereo pair. The relationship between estimate value and the frame number is graphed in Figure 3; the nominal value was measured to be 41.75 centimeters. Of course, the true separation is unmeasurable, so, as a proxy, the distance was measured between the respective corners of the cameras' cases. The second result, also shown in Figure 3, plots the coplanarity error over the frame number. The values for each wall were computed separately and appropriately averaged. The metric is defined as:

$$\frac{1}{n(n-1)} \sum_i \sum_{j \neq i} \left| \hat{\mathbf{z}}^T \mathbf{R_i^T} (\mathbf{t_j} - \mathbf{t_i}) \right|, \qquad \text{where } i, j \text{ range from 1 to } n.$$

This computes an average distance from each target's plane to every other target's origin; the result for perfectly coplanar targets is zero.

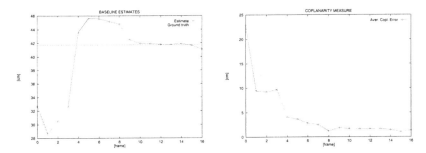

**Figure 3.** On the left are estimated baseline lengths during the reconstruction; the nominal value is displayed as well. On the right is the coplanarity error metric plotted over frame number.

# 5  Conclusion

The classic solutions of the static registration problem have been found insufficient for many augmented reality applications. By introducing a constraint on the data set — specifically planar targets — a highly accurate, metric reconstruction was obtained, as verified in qualitative and quantitative experiments. Given the simplicity of creation and placement of these targets, the presented technique provides a useful solution to the static registration problem.

# References

[Azuma 97]    R.T. Azuma. "A Survey of Augmented Reality," *Presence*, Vol. 6, pp. 355–385, 1997.

[Bajura 95]    M. Bajura and U. Neumann. "Dynamic Registration Correction in Video-Based Augmented Reality Systems," *IEEE Computer Graphics and Applications*, Vol. 15, No. 5, pp. 52–60, 1995.

[Huang 94]    T.S. Huang and A.N. Netraval. "Motion and Structure from Feature Correspondences: A Review," *Proceedings of the IEEE*, Vol. 82, pp. 251–268, 1994.

[Koller 97]    D. Koller, G. Klinker, E. Rose, D. Breen, R. Whitaker, and M. Tuceryan. "Automated Camera Calibration and 3D Egomotion Estimation for Augmented Reality Applications," *International Conference on Computer Analysis of Images and Patterns*, Springer Verlag, Berlin, pp. 199–206, 1997.

[Kumar 96]      R. Kumar and A.R. Hanson. "Robust Methods for Estimaging
                Pose and a Sensitivity Analysis," *Computer Vision and Image
                Understanding*, Vol. 60, pp. 313–342, 1994.

[Liu 90]        Y. Liu, T.S. Huang, and O.D. Faugeras. "Determination of Cam-
                era Location from 2-D to 3-D Line and Point Correspondences,"
                *IEEE Pattern Analysis and Machine Intelligence*, Vol. 12, pp.
                28–37, 1990.

[Maybank 92]    S.J. Maybank and O.D. Faugeras. "A Theory of Self-calibration
                of a Moving Camera," *International Journal of Comupter Vi-
                sion*, Vol. 8, pp. 123–151, 1992.

[Mellor 95]     J.P. Mellor. "Enhanced Reality Visualization in a Surgical En-
                vironment," Technical Report 1544, Massachusetts Institute of
                Technology Artificial Intelligence Laboratory, 1995.

[Oliensis 98]   J. Oliensis. "A Critique of Structure from Motion Algorithms,"
                Technical Report, NEC Research Institute, April 1997.

[Tsai 89]       R. Tsai. "Synopsis of Recent Progress on Camera Calibration for
                3D Machine Vision," In O. Khatib et al., editor, *The Robotics
                Review*, Cambridge, MA: MIT Press, pp. 147–159, 1989.

# Natural Feature Tracking for Extendible Robust Augmented Realities

Jun Park, Suya You, and Ulrich Neumann

**Abstract.** *Vision-based tracking systems for augmented reality often require that artificial fiducials be placed in the scene. In this paper, we utilize our approach for robust detection and tracking of natural features such as textures or corners. The tracked natural features are automatically calibrated to the fiducials that are used to initialize and facilitate normal tracking. Once calibrated, the natural features are used to extend the system's tracking range and to stabilize the tracked pose against occlusions and noise. The emphasis of this paper is the integration of natural feature tracking with fiducial tracking in order to increase the range and robustness of vision-based augmented reality tracking.*

## 1 Introduction

Many systems are designed to track the six degree-of-freedom (6DOF) pose of an object (or person) relative to a fixed-coordinate frame in its environment [Foxlin 98] [Ghazisadedy 95] [Meyer 92] [State 96] [Welch 97]. These tracking systems use a variety of sensing technologies, each with unique strengths and weaknesses, to measure a world-coordinate pose — required for virtual and augmented reality applications.

However, a large class of AR applications require annotation on objects whose position varies freely without any impact on the AR media linked to them (i.e., AR applications in manufacturing, maintenance, and training [Caudell 92] [Feiner 93]). A more appropriate tracking approach for these mobile applications is based on viewing the object itself [Neumann 96][Rekimoto 97] [Sharma 97] [Uenohara 95]. Many of these AR systems depend on artificial fiducials (also called landmarks) or prior-known model data to perform the dynamic alignment between a real and a virtual camera. These approaches are appropriate in situations where known and recognizable features are continually in view. However, it becomes difficult to ensure that this viewing constraint is satisfied in general application settings where off-screen fiducials, or occlusion of fiducials, are difficult to prevent. In these cases, continued-pose tracking depends on the utilization of all naturally occurring scene features. In this paper, we describe an approach for extending and stabilizing AR tracking by using natural features.

Our emphasis is upon the original combination of components and the architecture of our system. To the best of our knowledge, the idea of combining uncalibrated natural features with fiducials in an AR tracking system is novel.

## 2   Operation Model

When a user moves a camera that views a scene, and the system continuously tracks the fiducials as well as any natural features detected in the images, the camera pose (computed from the reference fiducials, i.e., by fiducial-based tracking) is used to calibrate natural scene features as they are tracked over an extended set of frames. Once calibrated, the natural scene features are used as tracking primitives to compute pose in the absence of reference fiducials (this is called natural-feature-based tracking). Using this hybrid approach, our system tracks robustly over a wider range of views and under conditions where reference fiducials are occluded.

Our system extends the use of a Kalman filter or a 3D-based recursive filter to compute structure from motion for calibrating tracked, natural features [Neumann 98b]. The detection and tracking of natural features is performed by a novel, adaptive, motion-estimation approach [Neumann 98a]. We select the most reliable natural features, based on a tracking confidence metric, and track these features in a multi-stage process that includes feedback on feature tracking confidence. The natural feature tracking is still an order of magnitude too slow ($\sim 1$ Hz) for real-time applications. For this paper, real-time video sequences are recorded and processed automatically (no user intervention) and then reassembled into a video sequence. We are currently investigating dedicated hardware and DSP processor implementations of the algorithms.

## 2.1   Fiducial-based Tracking

Fiducials have several advantages: they can be designed to maximize performance of AR systems to detect and distinguish between fiducials; they can be inexpensive; and they can be placed arbitrarily on objects. Our fiducial design is a colored circle or triangle [Neumann 96], but other designs such as concentric circles or coded squares are equally valid [State 96] [Cho 98]. Detection strategies are often dependent on the characteristics of the fiducials [Rekimoto 97] [Uenohara 95]. Currently, we are working on a more robust and intensity-tolerant fiducial detection system using fuzzy and rule-based algorithms [Cho 98].

Computing correspondences between 2D measurements and 3D features in a fiducial database is hard in the general case [Uenohara 95], and trivial if the fiducial types are unique for each element in the database. A small number of initial, reference fiducials are needed for this work of fiducial-based tracking; therefore, we used unique types of fiducials to solve the 2D-to-3D correspondence problem. In computing camera pose, three or more corresponding fiducials are needed [Fischler 81]. The approach we use has known instabilities in certain poses and, in general, provides multiple solutions (two or four) from a fourth-degree polynomial. Methods have been proposed for selecting the most likely solution [Sharma 97]. We utilize several tests to rank the possible pose solutions in terms of their apparent correctness [Neumann 98b].

## 2.2   Natural-Feature-Based Tracking

Our novel approach to natural-feature-based tracking is to combine natural feature tracking and new point-position estimation. This approach uses naturally occurring scene features to compute camera poses in unprepared environments.

The system detects natural features from the initial image and tracks in the subsequent images, while providing the image coordinates of the natural features for each image frame. These image coordinates of natural features, combined with the camera pose calculated by fiducial-based tracking, are used to calibrate the natural features using a new point-position estimation process.

The correspondence of natural features from one image to another is solved by a feature-tracking algorithm, as explained in Section 3. The estimated 3D positions and corresponding image coordinates of the natural features can be used to calculate the camera pose, if part or all of the fiducials are occluded or undetectable.

Our contribution is the unique combination of natural feature tracking and new point-position estimation to track the camera pose in an uncalibrated environment. Starting from a small set of calibrated fiducials, the camera tracking range can be extended dynamically and automatically. The user can also interact with the new environment, adding or modifying the virtual objects and real objects, since our tracking system is dynamically extendible to regions without calibrated features.

One of the issues related to natural-feature-based tracking is the strategy for selecting the natural features used to compute the camera pose.

Currently we have the strategy of dynamically choosing four features closest to each corner of the image. This results in evenly distributed feature sets. We also plan to take into consideration the uncertainties of 3D positions and the variances of 2D features.

## 3   Natural Feature Tracking

We developed a novel motion tracking approach to perform detection and tracking of naturally occurring features [Neumann 98a]. Our system integrates three main motion-analysis functions (e.g., feature selection, motion tracking, and estimate verification,) in a closed-loop, cooperative manner to deal with complex, natural-imaging conditions.

In the feature selection module, two kinds of tracking features (points and regions) are selected and evaluated for their ability to track and to provide motion estimation reliably. The selection and evaluation process is based on a tracking evaluation function that measures the suitability of features and the confidence in the tracking fed back from the tracking module. Once selected, the features are ranked according to their evaluation values, and then fed into the tracking module.

The tracking method we employed is a robust, multi-stage, optical-flow estimation approach [Neumann 98a]. For every region, an estimated motion is computed and fit into a model. Verification and evaluation are imposed to measure the confidence of the estimation and the model fit. If the estimation error is large, the motion estimate will be refined until the estimation error converges, or the region is discarded as unreliable for tracking. In order to handle local, geometric distortions due to large view variations and long sequence tracking, the translation model and the affine model are used to track point and region features, respectively. These models are the basis of the motion verification and evaluation processing. In some regions, for example, the optical-flow motion estimate is computed and fit to an affine model, then used to warp the regions into a confidence evaluation frame. To obtain a measure of tracking error, the confidence evaluation frame is then compared with the true target image.

Motion-error feedback is an essential component of the architecture for robust tracking. The error information is fed back to the tracking module for motion correction, and to the feature detection module for continuous feature re-evaluation. The re-evaluation of features keeps the tracking system working in an "optimum" state, by selecting and maintaining the most reliable features. Although two different verification strategies are used for the two types of tracked features (points and regions) and their corresponding motion models (translation and affine), they both generate evaluation frames that measure the estimation residual.

The closed-loop stabilization of the tracking system is inspired by the use of feedback for correcting errors in a non-linear control system. This process acts as a "selection-hypothesis-verification-correction" strategy that makes it possible to discriminate between good and poor estimation features, and maximizes the quality of the final motion estimation.

# 4   New Point-Position Estimation

Two different recursive filters have been designed and tested to estimate the 3D positions of new points based on the camera pose and measurements of feature image coordinates [Neumann 98b]. The results of synthetic and real data show that both filters converged and were stable. The EKF (Extended Kalman Filter) is known to have good characteristics under certain conditions; the RAC (Recursive Average of Covariances) filter gives comparable results, and it is simpler, operating completely in 3D-world space with 3D lines as measurements. The RAC filter approach replaces the linearization processes required by the EKF with Jacobian matrices. More details about EKF can be found in references such as [Maybeck 79] and [Mendel 95].

The intersection of two lines connecting the camera positions and the feature locations of the image creates the initial estimate of the 3D position of the feature.[1] The intersection threshold is a design parameter that depends on the fiducial design or the natural scene and objects. These initial estimates often have quite good accuracy, especially when the angles between the two lines are big enough (e.g., > 5 degrees).

# 5   Results and Discussion

Two models (a rack and a truck) were used in experiments. The image streams generated by a camera were directly digitized using a video editing system. The sampling rate was 15 Hz, resulting in about 250 frames for a 17- second video sequence of a rack model. For the truck model, the sampling rate was 25 Hz, resulting in about 340 frames of images for a 14-second video sequence. Figure 1 shows the result of natural feature tracking. In both rack and truck models, 20 features were detected in the first frame; 12 of these were selected for tracking, while the others were

---

[1]Since two lines in 3D space may not actually intersect, the point midway between the points of closest approach is used as the intersection.

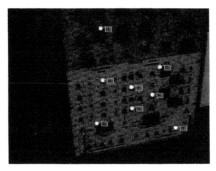 

Rack model: 250th frame          Truck model: 451st frame

**Figure 1.** Results of feature detection and tracking.

rejected as too close to fiducials or to the already-selected features. Even in later frames, all the features were accurately tracked except for those out of screen.

Figure 2 shows the results of new point-position estimation. Each chart marks the convergence of $X$, $Y$, and $Z$ coordinates of 3D positions of the natural features. They converged quickly (at about the 90th frame, i.e., in 6 seconds) and were stable after the convergence. It is remarkable that the initial estimates of $Z$ coordinates were less accurate than those of the $X$ and $Y$ coordinates, which would be intuitive.

Figure 3 shows the results of camera-pose tracking and virtual object/annotation overlay. The bigger, dark circles indicate the projections

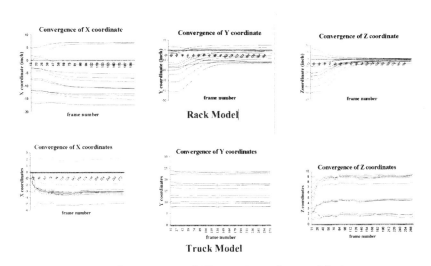

**Figure 2.** Convergence of 3D positions of natural features.

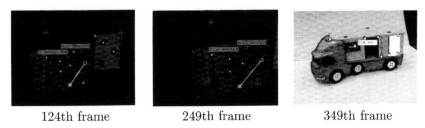

| 124th frame | 249th frame | 349th frame |

**Figure 3.** Result of virtual object/annotation overlay. (Left and middle) Rack model — virtual annotations and a virtual line. (Right) Truck model — a virtual annotation and a virtual white rectangle that covers a texture.

of the estimated 3D positions of the natural features. The smaller, bright circles indicate detected fiducials or the feature measurements that resulted from natural feature tracking. The bright crosses indicate fiducials or natural features that were used for tracking. In cases where fewer than three fiducials were detected, four features close to the four corners of the image (either tracked natural features or detected fiducials) were selected in order to compute the camera pose.

The experiments of the current work were done off-line, but automatically. Most of the computation time went to the natural feature tracking, which is still an order of magnitude too slow ($\sim$ 1 Hz) for real-time applications. It is hoped that dedicated hardware and DSP processor implementations of the algorithms will ultimately allow on-line and real-time applications. We plan to set up a strategy reflecting the uncertainties of 3D positions and 2D variances of the features in order to choose features with more accurate 3D positions and 2D image coordinates. We also hope to build a more robust pose calculation algorithm, which adapts multiple features and is more tolerant of numerical instability. A more detailed description of our work can be found in [Park 98].

# References

[Caudell 92]    T.P. Caudell and D.M. Mizell, "Augmented Reality: An Application of Heads-Up Display Technology to Manual Manufacturing Processes," *Proceedings of the Hawaii International Conference on Systems Sciences*, pp. 659–669, 1992.

[Cho 98]    Y.K. Cho, J.Lee, and U. Neumann, "A Multi-ring Color Fiducial System and a Rule-Based Detection Method for Scalable Fiducial-tracking Augmented Reality," *Proceedings of International Workshop on Augmented Reality (IWAR '98)*, A K Peters, Ltd., Natick, MA, pp. 147–165, 1999.

[Feiner 93]        S. Feiner, B. MacIntyre, and D. Seligmann, "Knowledge-Based Augmented Reality," *Communications of the ACM*, Vol. 36, No. 7, pp. 52–62, July 1993.

[Fischler 81]      M.A. Fischler and R.C. Bolles, "Random Sample Consensus: A Paradigm for Model Fitting with Applications to Image Analysis and Automated Cartography," *Graphics and Image Processing*, Vol. 24, No. 6, pp. 381–395, 1981.

[Foxlin 98]        E. Foxlin, M. Harrington, and G. Pfeifer, "Constellation: A Wide-Range Wireless Motion-Tracking System for Augmented Reality and Virtual Set Applications," *Proceedings of SIG-GRAPH '98, Computer Graphics*, pp. 371–378, 1998.

[Ghazisadedy 95]   M. Ghazisadedy, D. Adamczyk, D.J. Sandlin, R.V. Kenyon, and T.A. DeFanti, "Ultrasonic Calibration of a Magnetic Tracker in a Virtual Reality Space," *Proceedings of VRAIS '95*, IEEE Computer Society Press, Los Alamitos, CA, pp. 179–188, 1995.

[Maybeck 79]       P.S. Maybeck, *Stochastic Models, Estimation, and Control*, Volume 1, Acamedic Press, Inc., New York, NY, 1979.

[Mendel 95]        J.M. Mendel, *Lessons in Estimation Theory for Signal Processing, Communications, and Control*, Prentice Hall PTR, 1995.

[Meyer 92]         K. Meyer, H.L. Applewhite, and F.A. Biocca, "A Survey of Position Trackers," *Presence: Teleoperator and Virtual Environments*, Vol. 1, No. 2, pp. 173–200, 1992.

[Neumann 96]       U. Neumann and Y. Cho, "A Self-Tracking Augmented Reality System," *Proceedings of ACM Virtual Reality Software and Technology '96*, ACM, pp. 109–115, 1996.

[Neumann 98a]      U. Neumann and S. You, "Integration of Region Tracking and Optical Flow for Image Motion Estimation," *Proceedings of IEEE ICIP '98*, Chicago, IL, October 1998.

[Neumann 98b]      U. Neumann and J. Park, "Extendible Object-Centric Tracking for Augmented Reality," *IEEE Virtual Reality Annual International Symposium*, IEEE Computer Society Press, Los Alamitos, CA, 1998.

[Park 98]          J. Park and U. Neumann, "Extending Augmented Reality with Natural Feature Tracking," *Proceedings of SPIE, Telemanipulator and Telepresence Technologies V*, Vol. 3524, No. 15, Bellingham, November 1998.

[Rekimoto 97]    J. Rekimoto, "NaviCam: A Magnifying Glass Approach to Augmented Reality," *Presence: Teleoperator and Virtual Environments*, Vol. 6, No. 4, pp. 399–412, August 1997.

[Sharma 97]    R. Sharma and J. Molineros, "Computer Vision-Based Augmented Reality for Guiding Manual Assembly," *Presence: Teleoperator and Virtual Environments*, Vol. 6, No. 3, pp. 292–317, June 1997.

[State 96]    A. State, G. Hirota, D.T. Chen, B. Garrett, and M. Livingston, "Superior Augmented Reality Registration by Integrating Landmark Tracking and Magnetic Tracking," *Proceedings of SIGGRAPH '96, Computer Graphics*, pp. 429–438, 1996.

[Uenohara 95]    M. Uenohara and T. Kanade, "Vision-Based Object Registration for Real-Time Image Overlay," *Proceedings of Computer Vision, Virtual Reality, and Robotics in Medicine*, pp. 13–22, 1995.

[Welch 97]    G. Welch and G. Bishop, "SCAAT: Incremental Tracking with Incomplete Information," *Proceedings of SIGGRAPH '97, Computer Graphics*, pp. 333–344, 1997.

# Making Augmented Reality Work Outdoors Requires Hybrid Tracking

Ronald T. Azuma, Bruce R. Hoff, Howard E. Neely III,
Ronald Sarfaty, Michael J. Daily, Gary Bishop,
Leandra Vicci, Greg Welch, Ulrich Neumann,
Suya You, Rich Nichols, and Jim Cannon

**Abstract.** *Developing augmented reality systems that work outdoors, rather than indoors in constrained environments, will open new application areas and motivate the construction of new, more general tracking approaches. Accurate tracking outdoors is difficult because we have little control over the environment and fewer resources available than in an indoor application. This position paper examines the individual tracking technologies available and concludes that for the near term, a hybrid solution is the only viable approach. The distortion measured from an electronic compass and tilt sensor is discussed.*

## 1   Motivation

Several prototype augmented reality (AR) systems have demonstrated adequate performance to meet the needs of indoor applications, ranging from medical visualization [State 96] to aircraft manufacturing [Nash 97] and even entertainment [Ohshima 98]. However, few have attempted to build augmented reality systems that work outdoors. A group at Columbia University demonstrated the "touring machine," which allows a user to view information linked to specific buildings on the Columbia campus as he walks around outside [Feiner 97]. Some wearable computers, such as the CMU VuMan system, have been used for vehicle maintenance applications in outdoor settings. But none of these projects have attempted to achieve accurate registration at a wide variety of outdoor locations.

Augmented reality systems that provide accurate registration outdoors are of interest because they would make possible new application areas and could provide a natural interface for wearable computers, an area of growing interest both in academia and industry. A user walking outdoors could see spatially-located information directly displayed upon her view of the environment, helping her to navigate and identify features of interest. Today, a hiker in the woods needs to pull out a map, a compass, and a GPS (Global Positioning System) receiver; convert the GPS and compass readings to her location and orientation; and then mentally align the information from the two-dimensional map with what she sees in the three-dimensional environment around her. A personal, outdoor AR system could perform the same task automatically and display the trail path and landmark locations directly upon her view of the surrounding area, without burdening the hiker with the cognitive load.

Architectural clients could see what a proposed building would look like by walking around the construction site. Soldiers could see the locations of enemies, friends, and dangerous areas like minefields which might not be readily apparent to the naked eye. Personal, outdoor AR systems would also be useful for groups of people working together. If co-workers are widely separated, it is difficult for them to establish common frames of reference to describe spatially-located information. An instruction directing another team member to go to the "third white building to the left of the red building" may be useless if the recipient sees the world from a different vantage point than the speaker. Personal AR displays provide an unambiguous method of sharing such information.

Furthermore, personal outdoor AR displays might prove to be a natural interface for wearable PCs. The standard WIMP interface does not map well onto wearable PCs, because the desktop metaphor is not appropriate for a user walking around outdoors who might not have her hands free or be able to allocate complete attention to the computer [Rhodes 98]. Augmented reality might well be a better approach for certain applications.

What are the difficulties that prevent personal, outdoor AR systems from being deployed today? The ergonomic issues that face wearable PC systems apply to outdoor AR systems as well: developers must concern themselves with size, weight, power, ruggedness, etc. Displays that have sufficient contrast to work in outdoor settings are required. But the biggest challenge lies in accurate tracking outdoors — determining the user's position and orientation with sufficient accuracy to avoid significant registration errors. This position paper focuses on the tracking problem and discusses the issues involved.

Accurate tracking indoors is hard enough; accurate tracking outdoors is even more daunting because of two main differences in the situation. First, we have less control over the environment. Second, we have fewer resources available — power, computation, sensors, etc. These differences mean that solutions for indoor AR may not apply directly to personal outdoor AR systems. For example, several indoor AR systems have achieved accurate tracking and registration by carefully measuring the objects in a highly constrained environment, putting colored dots over those objects, and tracking the dots with a video camera (see Azuma 97 for some references). This approach violates some of the constraints of an outdoor situation. We do not have control over the outdoor environment and cannot always rely on modifying it to fit the needs of the system. For example, in a military application it is not realistic to ask soldiers, friendly or enemy, to wear large, brightly colored dots to aid our tracking system. We should not expect to measure every object in the environment beforehand. Also, this approach may require more computational resources than is practical for a single outdoor user. Such systems have used an SGI Onyx or other high-end workstation with frame-grabbing capability. However, outdoor AR applications may demand less registration accuracy than many indoor applications. A doctor performing a needle biopsy requires that the virtual incision marker be accurate within a millimeter, but for a hiker walking around, perhaps even one degree of angular error is acceptable when cuing the user to landmarks in the environment.

## 2 Analysis

If we analyze the tracking technologies available outdoors, we find that no single technology provides a complete solution. Combining several tracking technologies, or hybrid tracking, is the only feasible approach for the near term. Hybrid approaches increase system complexity and cost, but often provide the most robust results. We now briefly describe the strengths and weaknesses of individual technologies, including an analysis of data taken from a sourceless orientation sensor.

- GPS: The Global Positioning System provides worldwide coverage and typically measures the user's three-dimensional position within 30 meters for regular GPS and about 3 meters for differential. It does not measure orientation. Differential accuracy is sufficient for viewing distant but not nearby objects; at 50 meters range a 3-meter position error results in 3.4 degrees of registration error. New carrier-phase GPS receivers claim accuracy to within centimeters, but multipath problems make that difficult to achieve in many outdoor situations. GPS requires direct line-of-sight to the satellites and is commonly blocked in urban areas, canyons, etc. In military situations, GPS is easily jammed.

- Inertial and dead reckoning: Inertial sensors are sourceless and relatively immune to environmental disturbances. Their main problem is drift; they accumulate error with time, and existing inertial sensors of the cost and weight appropriate for a single person drift too quickly to be the sole solution for an outdoor system. Using accelerometers to track position is especially difficult due to the double integration required; this double integration requires accelerometers with bias accuracies several orders of magnitude better than what is commonly available today.

- Active sources: Users commonly set up active transmitters and receivers (using magnetic, optical, or ultrasonic technologies) for indoor Virtual Environment systems, but modifying an outdoor environment in this manner is not usually practical and restricts the user to the location of the active sources.

- Passive optical: The user's view of the surrounding environment provides information that can be used to extract the user location. Video sensors observe detectable features in the environment, such as the sun, the stars, or other recognizable objects. Such video-based tracking techniques enable closed-loop tracking approaches, an important capability for future outdoor AR systems. Video sensors are line-of-sight and will lose lock if the view is obscured (by buildings, vegetation, etc.). Computer vision techniques are not currently robust enough to provide a complete solution, and unless known landmarks are used, the tracking solutions are relative rather than absolute. Video processing is also computationally intensive and not feasible in real time on most wearable PC systems.

- Electronic compass and tilt sensors: Many inexpensive HMDs use a sourceless orientation-only head tracker that consists of an electronic compass and two tilt sensors. The tracker is small and inexpensive and claims yaw accuracy of 0.5 degrees. The compass is highly sensitive to disturbances in the ambient magnetic field. Figure 1 shows the equipment we used to measure distortions in the field. We built this mechanical turntable out of Delrin to avoid adding any sources of magnetic distortion. The measurements were taken outdoors at several different locations, far from any apparent sources of distortion. Even in such ideal environments, the compass measurements would be in error by up to $2 - 3$ degrees, and the distortion pattern varies significantly with time and location (Figure 2). The sensors' output is also delayed in time by roughly 100 milliseconds due to the

**Figure 1.** Non-metallic turntable for measuring compass distortion.

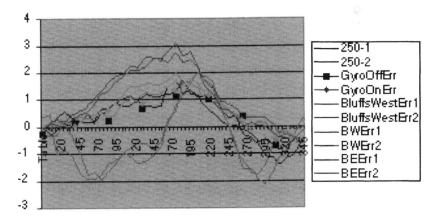

**Figure 2.** Distortion in electronic compass (Precision Navigation TCM2). *X* axis is yaw orientation of compass in degrees, *Y* axis is the relative error in degrees from the zero degree initial orientation.

settling time of the tilt sensors, and the output is noisy and displays noticeable overshoots during motion. Near metal objects, the distortion can increase to 20 or 30 degrees. While this sensor may be sufficient for some games and is helpful as part of a hybrid solution, our measurements strongly suggest that by itself such a tracker is inadequate for outdoor AR.

Since no single tracking technology appears adequate to provide a complete solution, the authors have formed a team to collaborate on research into hybrid tracking for outdoors AR. The team expects to combine the technologies listed above with others yet to be developed. This work is funded by DARPA under the GRIDS (Geospatial Registration of Information for Dismounted Soldiers) project.

## Acknowledgments

Work in this area has largely been funded by DARPA contract N00019-97-C-2013.

## References

[Azuma 97]    R.T. Azuma. "A Survey of Augmented Reality." *Presence: Teleoperators and Virtual Environments* Vol. 6, No. 4, pp. 355–385, August 1997.

[Feiner 97]    S. Feiner, B. MacIntyre, and T. Höllerer. "A Touring Machine: Prototyping 3D Mobile Augmented Reality Systems for Exploring the Urban Environment." *Proceedings of First International Symposium on Wearable Computers,* Cambridge, MA, pp. 74–81, October 1997.

[Nash 97]    J. Nash. "Wiring the Jet Set." *Wired* Vol. 5, No. 10, pp. 128–135, October 1997.

[Ohshima 98]    T. Ohshima, K. Satoh, H. Yamamoto, and H. Tamura. "AR2 Hockey: A Case Study of Collaborative Augmented Reality." *Proceedings of IEEE VRAIS '98,* Atlanta, GA, pp. 268–275, March 1998.

[Rhodes 98]    B. Rhodes. "WIMP Interface Considered Fatal." Position paper at *IEEE VRAIS '98 Workshop on Interfaces for Wearable Computers,* Atlanta, GA, March 1998.

[State 96]    A. State, M.A. Livingston, G. Hirota, W.F. Garrett, M.C. Whitton, H. Fuchs, and E.D. Pisano. "Techniques for Augmented- Reality Systems: Realizing Ultrasound-Guided Needle Biopsies." *Proceedings of SIGGRAPH '96,* New Orleans, LA, pp. 439–446, August 1996.

# Improving Registration Precision Through Visual Horizon Silhouette Matching

Reinhold Behringer

**Abstract.** *The registration precision of an augmented reality (AR) system for enhancing the situational awareness in an outdoor setting can be improved by utilizing visual clues. Terrain silhouettes can provide unique features to be matched with digital elevation map (DEM) data. The best match of a visually extracted silhouette with the DEM silhouette provides camera/observer orientation (elevation and azimuth angle). We have developed such a registration system which runs on a PC (Pentium Pro, 200 MHz) and is being ported to a wearable AR system.*

## 1 Introduction and Context

### 1.1 Augmented Reality

In the past several years, augmented reality (AR) has gained significant attention due to rapid progress in several key areas (e.g., wearable computing and virtual reality rendering) [Caudell 94]. AR technology provides an intuitive way to present information in order to enhance situational awareness and perception by exploiting the natural and familiar modes of human interaction with the environment. Completely immersive AR requires the user to wear head-mounted, see-through displays that require precise calibration [Janin 93].

### 1.2 Registration for AR

A very important issue in AR is the registration (alignment) between information to be displayed and the real world. Outdoor AR applications can utilize GPS for positioning, and magnetic compasses (magnetometers) and inclinometers for orientation, but these sensors lack precision. Inertial sensors can provide data during rapid motion, but they tend to drift. Hybrid tracking systems provide more promising results by combining inertial tracking with static positioning measurements and computer vision components.

Many AR systems employ computer vision technology for achieving required registration precision (see, e.g., Koller 97). The approach in most cases is to detect visual features that have a known position, and then to recover camera orientation and position through a matching or pose recovery process. Indoor AR applications often use fiducial markers for visual tracking [Hoff 96] [Neumann 96]. Visual landmark tracking, combined with magnetic tracking and inclinometers, can leverage the benefits of both tracking types in a hybrid system [State 96]. Correct camera calibration is crucial for many computer vision approaches, but methods have been developed to avoid explicit camera calibration [Kutulakos 96]. In general, AR benefits significantly from progress in computer vision research.

## 1.3   Silhouettes as Visual Clues

Contour silhouettes provide highly visible cues for the reconstruction of the shapes of curved objects [Vijayakumar 96] as well as for simultaneous motion estimation [Joshi 95]. In an outdoor scenario, there are silhouettes of man-made structures (buildings) as well as the natural shape of the terrain. If object and background have a significant intensity or color difference, the contour silhouettes can be detected by applying standard computer vision methods (edge detection and contour following).

Natural horizon silhouettes occur at the visual boundary between the terrain and the sky. Although not necessarily unique, they can be used to determine the position and orientation of the observer (the so-called "drop-off problem" [Thompson 90]), if the area is confined to a given region. Stein [Stein 92] demonstrates this process using digital elevation maps to match visual and predicted horizon silhouettes to structural indexing.

This paper describes an outdoor registration approach that is being developed by the Rockwell Science Center, using visual horizon silhouettes to improve the registration precision in an outdoor scenario. This approach can determine azimuth and elevation as well as camera scaling parameters.

## 2   About Digital Terrain Maps

Digital elevation maps provide terrain elevation data in a point grid pattern. The most common file formats are Digital Elevation Model (DEM) maps and Digital Terrain Elevation Data (DTED) maps. DEM data is available from the US Geological Survey (USGS) [USGSb]; DTED maps are available from the National Imagery and Mapping Agency [NIMA]. DEM maps are available in two formats: 1-degree and 7.5-minute maps. The elevation data in digital elevation maps are aligned in profile lines. Details about the file formats are published by USGS [USGSa]. The 1-degree format covers elevation data in a region of 1-deg. by 1-deg. The grid points are aligned to the geographic coordinate system along the latitude $\phi$ and longitude $\psi$ axes in a 3 seconds of arc raster. The 7.5-minute maps cover a region of 7.5 by 7.5 minutes of arc. The boundaries are the latitude and longitude axes, but the grid of the data sample points lies along the axes of the corresponding Universal Transverse Mercator (UTM) zone. A silhouette contour of a terrain is formed by those terrain points which have a surface normal vector perpendicular to the viewing vector, as seen from the observation point $P_0$ (Figure 1).

**Figure 1.** Normal and viewing vectors (left). Various silhouette points as seen from $P_0$ (right).

One way to compute silhouette contours from a digital elevation map is to calculate the scalar product of the vector normal $\vec{n_{ij}}$ and the viewing vector $\vec{v_{ij}}$ at each terrain point $P_{ij}$ and test it for equivalence to zero (or for lying between neighboring points where this scalar product has opposite signs). A simpler method is to select the points with the highest elevation visible from $P_0$.

## 3   Concept for Registration

Registration of the virtual world with the real world can be achieved by trying to match real-world features with predicted features from a known database. We have developed an approach for using DEM maps as a database for obtaining the camera orientation in a well-structured terrain by visual horizon silhouette matching. The position is assumed to be known through GPS measurements. The approach works as follows: First, the visual silhouette is extracted from a single video image by using edge-detection techniques. Next, extrema (peaks and dips) are extracted. Based on the known location of the camera/observer, the horizon silhouette for all 360 degrees of the surroundings is computed from a digital elevation map. Hypotheses are generated for the correspondence of the singularities.

These hypotheses define pitch and yaw angle, and camera calibration parameters. The hypothesis with the minimal distance between the visual silhouette segment and the silhouette from the digital elevation maps is then assumed to be the optimal match. It is then fine-tuned by further minimizing of the matching discrepancies.

The horizon silhouette in a video image is usually characterized by a steep, gray-scale gradient. This edge can be detected by applying an edge detector and scanning from the top to the bottom of the image. In our approach, a Sobel operator is used to detect the mainly horizontal edge. The occlusion of the horizon is problematic; if nearby trees or buildings cover the horizon edge, the horizon cannot be detected. Therefore, constraints must be strictly applied to flag areas where a horizon edge cannot be identified.

Using the horizon silhouette as visual cue for registration, the problem of determining the camera orientation is reduced to finding the optimal match of two arrays (visual silhouette segments and digitally predicted silhouette).

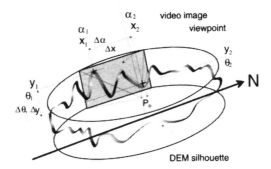

**Figure 2.** The generation of a hypothesis by assuming correspondence between one maximum and one minimum in both video and DEM silhouette.

Starting with a combination of two extrema in the video silhouette at $(x_1, y_1)$ and $(x_2, y_2)$, a hypothesis is generated for the corresponding extrema, $(\alpha_1, \theta_1)$ and $(\alpha_2, \theta_2)$, in the 360-degree predicted silhouette (Figure 2). The hypothetical match between these two pairs leads to the assumed camera azimuth $\alpha_c$ and elevation $\theta_c$. Calibration is not required; in fact, the matching hypothesis also provides a straight forward computation of the calibration parameters under the assumption of pinhole mapping.

After the hypothesis has been generated, an error measure is calculated to describe the error of the hypothesis. In order to compute the "distance" between the video and the predicted silhouette for each DEM silhouette point, the closest interpolated video-silhouette point is selected as a corresponding point. The square sum of all these distances $d_i$, within the field of view of the camera, is used as an error measure $e$ for judging the hypothesis. The hypothesis with the least error $e$ — below a given threshold — is chosen as the valid one.

## 4    Experimental Results

The approach to visual registration described in Section 3 has been implemented in a development setup for eventual integration into an outdoor AR system which provides visual aid for navigation and hazard indication. This system is currently being ported for operation in an outdoor vehicle, using the registration approach described in this paper along with other sensors (inertial, inclinometer, GPS).

For the development stage, the registration system was implemented on a 200-MHz Pentium Pro PC running Windows NT. The initial images were grabbed with a Cohu 2200 CCD camera (1/2" chip), using an image size of 512×480 pixels. The software is written in C++, using Microsoft Foundation Classes.

The video image in Figure 3 was captured by a black and white camera, at 512×480 pixels. It provides clear contrast between sky and terrain area and is used to demonstrate the registration approach. The silhouette extracted from this image is shown in Figure 4, together with the DEM silhouette segment.

In the DEM silhouette computed for the above location, there are 60 maxima and 60 minima. Combining these with the 10 maxima and 10 minima of the video silhouette results in 324,000 theoretical combinations for a correspondence hypothesis. After applying the constraints to the selection of possible combinations, only 75 combinations remain to be investigated.

**Figure 3.** Video image taken from the Rockwell Science Center near Thousand Oaks (CA). Top and bottom image portions have been truncated.

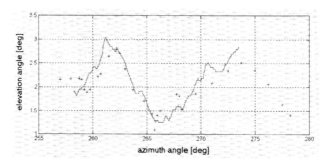

**Figure 4.** Best match of silhouettes. The contiguous line is the video silhouette, the dots are DEM silhouette points.


Table 1 lists the numerical results of the best match. The slight correction in the pitch angle is due to fine-tuning of the match by further minimization of the error measure.

|         | hypothesis | corrected  | error  |
| ------- | ---------- | ---------- | ------ |
| $\alpha_c$ | 265.9°     | 265.9°     | ±0.3°  |
| $\theta_c$ | 1.54°      | 1.62°      | ±0.2°  |
| $C_x$   | 33.1 pix/° | 33.3 pix/° |        |
| $C_y$   | 21.3 pix/° | 21.2 pix/° |        |

**Table 1.** Matching results.

# 5   Summary and Conclusion

We have demonstrated that in a well-structured terrain, horizon silhouettes can provide visual clues to improve registration. Our approach provides means for obtaining camera azimuth and elevation by using a non-calibrated camera; it even provides the calibration parameters. Further work needs to be done to make this approach applicable under a wider variety of lighting and visibility conditions. Principal limitations of computer vision in outdoor scenarios, however, require that a robust registration system include other sensors.

# Acknowledgments

This research was funded by the FedLab Advanced Displays and Interactive Displays Consortium, Cooperative Agreement DAAL01-96-2-0003.

# References

[Caudell 94]     T.P. Caudell. "Introduction to Augmented and Virtual Reality." *Proceedings of SPIE Conference on Telemanipulator and Telepresence Technologies,* Boston, MA, pp. 272–281, October 1994.

[Hoff 96]        W.A. Hoff and K. Ngyen. "Computer Vision-based Registration Techniques for Augmented Reality." *Proceedings of SPIE '96,* Boston, MA, pp. 538–548, November 1996.

[Janin 93]       A.L. Janin, D. Mizell, and T. Caudell. "Calibration of Head-Mounted Displays for Augmented Reality Applications." *Proceedings of VRAIS '93,* Seattle, WA, pp. 246–55, September 1993.

[Joshi 95]       T. Joshi, N. Ahuja, and J. Ponce. "Structure and Motion Estimation from Dynamic Silhouettes Under Perspective Projection." *Proceedings of ICCV '95,* Cambridge, MA, June 1995.

[Koller 97]      D. Koller, G. Klinker, E. Rose, D. Breen, R. Whitaker, and M. Tuceryan. "Real-time Vision-based Camera Tracking for Augmented Reality Applications." *Proceedings of VRST '97,* Lausanne, Switzerland, September 1997.

[Kutulakos 96]   K.N. Kutulakos and J. Vallino. "Affine Object Representations for Calibration-free Augmented Reality." *Proceedings of VRAIS '96,* Santa Clara, CA, pp. 25–36, 1996.

[Neumann 96]     U. Neumann and Y. Cho. "A Self-tracking Augmented Reality System." *Proceedings of VRST '96,* Hong Kong, July 1996.

[NIMA]           National Imagery and Mapping Agency web site. http://www.nima.mil/

[State 96]       A. State, G. Hirota, D.T. Chen, W.F. Garrett, and M.A. Livingston. "Superior Augmented Reality Registration by Integrating Landmark Tracking and Magnetic Tracking." *Proceedings of SIGGRAPH,* New Orleans, LA, 1996.

[Stein 92]       F.J. Stein. *Structural Indexing for Object Recognition.* PhD dissertation, University of Southern California, April 1992.

[Thompson 90]    W.B. Thompson, H.L. Pick, B.H. Bennet, M.R. Heinrichs, S.L. Savitt, and K. Smith. "Map-based Localization: The Drop-off Problem." *Proceedings of DARPA Image Understanding Workshop,* Pittsburgh, PA, September 1990.

[USGSa]          U.S. Geological Survey. *Digital Elevation Models — Data Users Guide 5,* United States Department of the Interior, Reston, VA, 1993.

[USGSb]              United States Geological Survey web site. http://www-nmd.
                     usgs.gov/

[Vijayakumar 96]     B. Vijayakumar. *Recognition and Pose Estimation of 3D
                     Curved Objects with Models Built Using Active Vision.* PhD
                     dissertation, Yale University, May 1996.

# Contributor List

**Ronald Azuma**; HRL Laboratories, 3011 Malibu Canyon Road MS RL96, Malibu, CA 90265-4799; Phone 310-317-5151; Fax 310-317-5695; Email azuma@hrl.com

**Ruzena Bajcsy**; Computer & Information Science, University of Pennsylvania, 3401 Walnut Street, Lab #303C, Philadelphia, PA 19104-6228; Email bajcsy@central.cis.upenn.edu

**Reinhold Behringer**; Rockwell Science Center, 1049 Camino Dos Rios, Thousand Oaks, CA 91360; Phone 805-373-4435; Fax 805-373-4862; Email reinhold@rsc.rockwell.com

**Jeffrey Berger**; Computer Vision Laboratory, Scheie Eye Institute, University of Pennsylvania 51 North 39th Street, Philadelphia, PA 19104; Phone 215-662-8675; Fax 215-662-0133; Email jwberger@mail.med.penn.edu

**Marie-Odile Berger**; LORIA, BP 239, 54506 Vandoeuvre-les-Nancy, France; Phone ++33-3-83-59-2067; Email berger@loria.fr

**Gary Bishop**; University of North Carolina at Chapel Hill, CB #3175 Computer Science Department, Chapel Hill, NC 27599-3175; Phone 919-962-1886; Fax 919-962-1799; Email gb@cs.unc.edu

**Jim Cannon**; Raytheon Systems Company, 2175 Park Place M/S A264, El Segundo, CA 90245; Phone 310-616-3401; Fax 310-616-4729; Email jjcannon@west.raytheon.com

**Youngkwan Cho**; Computer Science Department, IMSC, University of Southern California, 3740 McClintock Ave. Ste 131, Los Angeles, CA 90089-2561; Phone 213-740-3816; Fax 213-740-5807; Email ykcho@usc.edu

**Wendell Curtis**; The Boeing Company, 10907 165th Pl. NE, Redmond, WA 98052; Phone 425-865-3731; Fax 425-865-2966; Email wendell.d.curtis@boeing.com

**Kostantinos Daniilidis**; University of Pennsylvania, 3401 Walnut Street, GRASP Lab #336C, Philadelphia, PA 19104-6228; Phone 215-898-8549; Fax 215-573-2048; Email mendels@seas.upenn.edu

**Michael Daily**; HRL Laboratories, 3011 Malibu Canyon Road MS RL96, Malibu, CA 90265; Phone 310-317-5673; Fax 310-317-5695; Email mjdaily@hrl.com

**Anne-Laure Fayard**; INSEAD, Technology Management Department, Boulevard de Constance, 77305 Fontainebleau cedex, France; Email anne-laure.fayard@der.edfgdf.fr

**Henry Fuchs**; Department of Computer Science, University of North Carolina, 209 Sitterson Hall, Chapel Hill, NC 27599-3175; Phone 919-962-1911; Fax 919-962-1799; Email fuchs@cs.unc.edu

**Peter Gruenbaum**; Linius Technologies, 276 Turnpike Road, Westboro, MA 01581; Phone 508-626-9370; Fax 508-616-9362; Email pgruenbaum@linius.com

**Bruce Hoff**; HRL Laboratories, 3011 Malibu Canyon Road MS RL96, Malibu, CA 90265; Phone 310-317-5674; Fax 310-317-5695; Email hoff@hrl.com

**William Hoff**; Engineering Division, Colorado School of Mines, Golden, CO 80401; Phone 303-273-3761; Fax 303-273-3602; Email whoff@mines.edu

**Adam Janin**; International Computer Science Institute, 1947 Center Street, Berkeley, CA 94708; Phone 510-642-4274 x177; Fax 501-643-7684; Email janin@cs.berkeley.edu

**Tony Jebara**; MIT Media Lab, E15-390, 20 Ames Street, Cambridge, MA 02139-4307; Phone 617-253-0326; Fax 617-253-8874; Email jebara@media.mit.edu

**Gudrun Klinker**; Moos 2, D-85614 Kirchseeon, Germany; Phone ++49-89-289-25784; Fax ++49-89-289-25296; Email: klinker@in.tum.de

**Shaun Lawson**; Mechtronic Systems & Robotics Research, University of Surrey, Guildford, Surrey GUZ 5XH, U.K.; Phone ++44-1483-259681; Email s.lawson@surrey.ac.uk

**Jongweon Lee**; Computer Science Department, IMSC, University of Southern California, 3740 McClintock Ave. Ste 131, Los Angeles, CA 90089-2561; Phone 213-740-3816; Fax 213-740-5807; Email jonlee@usc.edu

**Vincent Lepetit**; LORIA, BP 239, 54506 Vandoeuvre-les-Nancy, France; Phone ++33-3-83-59-2067; Email lepetit@loria.fr

**Wendy Mackay**; Department of Computer Science, Aarhus University, Aabogade 34, DK-8200, Aarhus N, Denmark; Phone ++45-89-42-5622; Fax ++45-89-42-5624; Email mackay@daimi.au.dk

**Jeffrey Mendelsohn**; University of Pennsylvania, 3401 Walnut Street, GRASP Lab #336C, Philadelphia, PA 19104-6228; Phone 215-898-8549; Fax 215-573-2048; Email mendels@seas.upenn.edu

**David Mizell**; The Boeing Company, P.O. Box 3707, Mailstop 7L-48, Seattle, WA 98124-2207; Phone 425-865-2705; Fax 425-865-2965; Email david.mizell@boeing.com

**Jose Molineros**; Department of Computer Science and Engineering, Pennsylvania State University, 220 Pond Lab, University Park, PA 16902-6106; Phone 814-867-3786; Email josesito@psu.edu

**Stefan Müller**; Fraunhofer IGD, Department Visualization and Virtual Reality, Rundeturmstraße 6, D-64283 Darmstadt, Germany; Phone ++49-6151-155 188; Fax ++49-6151-155196; Email stefanm@igd.fhg.de

**Howard Neely III**; HRL Laboratories; 3011 Malibu Canyon Road MS RL96, Malibu, CA 90265; Phone 310-317-5829; Fax 310-317-5695; Email neely@hrl.com

**Ulrich Neumann**; Computer Science Department, Integrated Media Systems Center, University of Southern California, 3740 McClintock Ave. Ste 131, Los Angeles, CA 90089-2561; Phone 213-740-4486; Fax 213-740-8931; Email uneumann@usc.edu

**Rich Nichols**; Raytheon Systems Company, 2175 Park Place M/S A263, El Segundo, CA 90245; Phone 310-616-4999; Fax 310-616-4729; Email rwnichols@west.raytheon.com

**Oliver Nuria**; MIT, 20 Ames Street, Cambridge, MA 02139-4307; Phone 617-253-0326; Fax 617-253-8874; Email nuria@media.mit.edu

**Toshikazu Ohshima**; Mixed Reality Systems Laboratory Inc.,
6-145 Hanasaki-cho, Nishi-ku, Yokohama 220-0022, Japan;
Phone ++81-45-4118111; Fax ++81-45-4118110; Email ohshima@mr-system.com

**Jun Park**; Computer Science Department, Integrated Media Systems Center,
University of Southern California, 3740 McClintock Ave. Ste 131, Los Angeles,
CA 90089-2561; Phone 213-740-3816; Fax 213-740-5807;
Email junp@tracker.usc.edu

**Alex Pentland**; MIT, 20 Ames Street, Cambridge, MA 02139;
Phone 617-253-0370; Fax 617-253-8874; Email sandy@media.mit.edu

**John Pretlove**; Information Technology and Control Systems, ABB Corporate
Research, Bergerveien 12, 1375 Billingstad Norway; Phone ++44-1483-300800;
Email j.pretlove@surrey.ac.uk

**Vijaimukund Raghavan**; Department of Computer Science and Engineering,
Pennsylvania State University, 220 Pond Lab, University Park, PA 16902-6106

**Ramesh Raskar**; University of North Carolina, Sitterson Hall,
CB #3175 Computer Science Department, Chapel Hill, NC 27599-3175;
Phone 919-962-1761; Fax 919-962-1799; Email raskar@cs.unc.edu

**Dirk Reiners**; Fraunhofer IGD, Department Visualization and Virtual Reality,
Rundeturmstraße 6, D-64283 Darmstadt, Germany; Phone ++49-6151-155 273;
Fax ++49-6151-155196; Email reiners@igd.fhg.de

**Bradley Rhodes**; MIT Media Lab, E15-305D, 20 Ames Street, Cambridge, MA
02139-4307; Phone 617-253-9601; Fax 617-253-6215; Email rhodes@media.mit.edu

**Ronald Sarfaty**; HRL Laboratories, 3011 Malibu Canyon Road MS RL72,
Malibu, CA 90265; Phone 310-317-5013; Fax 310-317-5695;
Email rsarfaty@hrl.com

**Kiyohide Satoh**; Mixed Reality Systems Laboratory Inc., 6-145 Hanasaki-cho,
Nishi-ku, Yokohama 220-0022, Japan; Phone ++81-45-4118111;
Fax ++81-45-4118110; Email ksato@mr-system.co.jp

**Bernt Schiele**; MIT Media Lab, E15-384C, 20 Ames Street, Cambridge, MA
02139; Phone 617-253-0370; Fax 617-253-8874; Email bernt@media.mit.edu

**Rajeev Sharma**; Department of Computer Science and Engineering,
Pennsylvania State University, 220 Pond Lab, University Park, PA 16902-6106;
Phone 814-867-0147; Fax 814-865-3176; Email rsharma@cse.psu.edu

**David Shin**; University of Pennsylvania, Computer Vision Laboratory,
Scheie Eye, 51 North 39th Street, Philadelphia, PA 19104; Phone 215-662-8675;
Fax 215-662-0133; Email dsshin@mail.med.upenn.edu

**Gilles Simon**; LORIA, BP 239, 54506 Vandoeuvre-les-Nancy, France;
Phone ++33-3-83-59-2067; Email gsimon@loria.fr

**Thad Starner**; Georgia Institute of Technology College of Computing,
801 Atlantic Drive, Atlanta, GA 30332-0280; Phone 404-894-0816;
Fax 404-894-0673; Email thad@cc.gatech.edu

**Didier Stricker**; Fraunhofer IGD, Department Visualization and Virtual Reality, Rundeturmstraße 6, D-64283 Darmstadt, Germany; Phone ++49-6151-155
275; Fax ++49-6151-155196; Email stricker@igd.fhg.de

**Venkataraman Sundareswaran**; Rockwell Science Center, 1049 Camino Dos Rios, Thousand Oaks, CA 91360; Phone 805-373-4845; Fax 805-373-4862; Email vsundar@rsc.rockwell.com

**Hideyuki Tamura**; Mixed Reality Systems Laboratory Inc.,
6-145 Hanasaki-cho, Nishi-ku, Yokohama 220-0022, Japan;
Phone ++81-45-4118111; Fax ++81-45-4118110; Email tamura@mr-system.com

**Jim Vallino**; Department of Computer Science, Rochester Institute of Technology, 102 Lomb Memorial Drive, Rochester, NY 14623-5608; Phone 716-475-2991; Fax 716-475-7100; Email jrv@cs.rit.edu

**Leandra Vicci**; University of North Carolina at Chapel Hill,
CB #3175 Computer Science Department, Chapel Hill, NC 27599-3175;
Phone 919-962-1742; Fax 919-962-1799; Email chi@cs.unc.edu

**Joshua Weaver**; MIT, E15, 20 Ames Street, Cambridge, MA 02139;
Phone 617-253-0370; Fax 617-253-8874; Email joshw@media.mit.edu

**Greg Welch**; University of North Carolina, Sitterson Hall, CB #3175 Computer Science Department, Chapel Hill, NC 27599-3175; Phone 919-962-1819;
Fax 919-962-1799; Email welch@cs.unc.edu

**Hiroyuki Yamamoto**; Mixed Reality Systems Laboratory Inc., 6-145 Hanasaki-cho, Nishi-ku, Yokohama 220-0022, Japan; Phone ++81-45-4118111;
Fax ++81-45-4118110; Email ymmt@mr-system.com

**Suya You**; Computer Science Department, IMSC, University of Southern California, 3740 McClintock Ave. Ste 131, Los Angeles, CA 90089-2561;
Phone 213-740-4495; Fax 213-740-5807; Email suyay@usc.edu

Printed and bound by CPI Group (UK) Ltd, Croydon, CR0 4YY

23/10/2024

01777671-0002